THE YOUNG OXFORD BOOK OF

Ecology

Ecology

Michael Scott

Oxford University Press

Oxford University Press

Oxford New York
Athens Auckland Bangkok Bombay
Calcutta Cape Town Dar es Salaam Delhi
Florence Hong Kong Istanbul Karachi
Kuala Lumpur Madras Madrid Melbourne
Mexico City Nairobi Paris Singapore
Taipei Tokyo Toronto Warsaw

and associated companies in
Berlin Ibadan

Copyright © 1995 by Michael Scott
First published in paperback 1998

Published by Oxford University Press, Inc.
198 Madison Avenue
New York, New York 10016

Oxford is a registered trademark of Oxford
University Press

Library of Congress Cataloging-in-Publication Data

Scott, Michael (Michael M.)
Ecology/ Michael Scott.
p. cm.—(Young Oxford books)
Includes index.
1. Ecology—Juvenile literature.
[1. Ecology. 2. Habitat (Ecology)]
I. Title. II. Series.
QH541.14.S385   1995
574.5—dc20
95-7012
CIP
AC

ISBN 0-19-521428-5 (paper ed.)
ISBN 0-19-521166-9 (lib. ed.)

9 8 7 6 5 4 3 2 1

Printed in Italy by
G. Canale & C. S.p.A. - Borgaro Tse (Turin)

Front cover A bee pollinating a flower.

CONTENTS

1

LIFE-STYLES

2

LIFE ON EARTH

3

THREATS TO LIFE

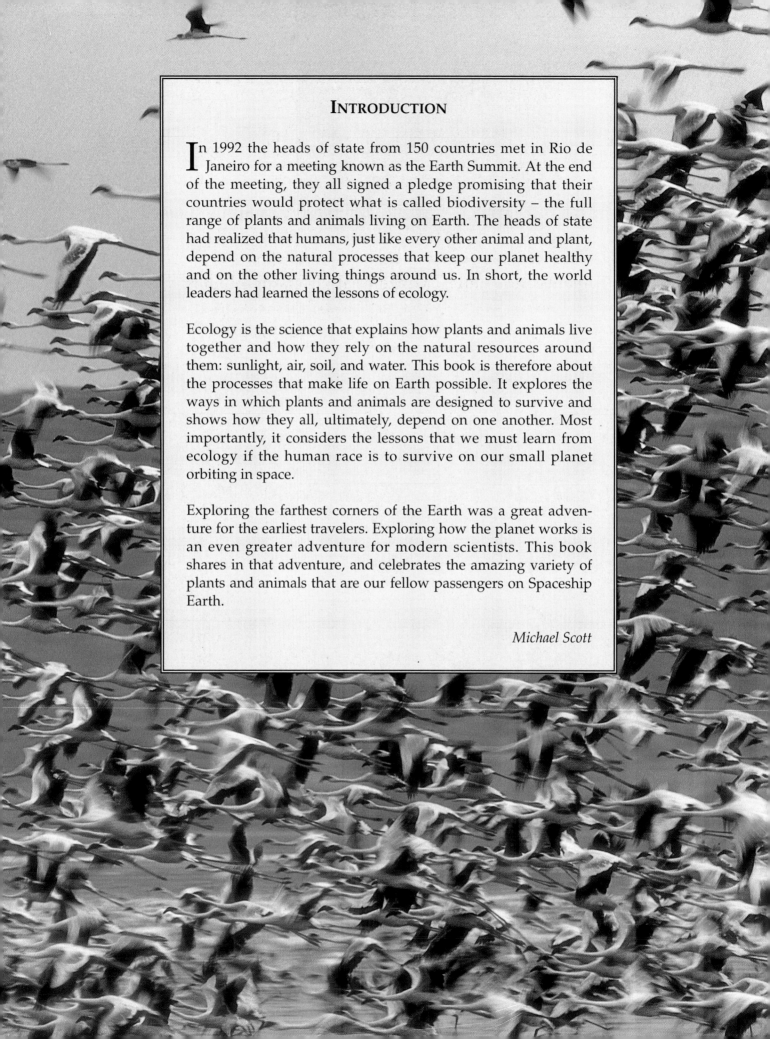

INTRODUCTION

In 1992 the heads of state from 150 countries met in Rio de Janeiro for a meeting known as the Earth Summit. At the end of the meeting, they all signed a pledge promising that their countries would protect what is called biodiversity – the full range of plants and animals living on Earth. The heads of state had realized that humans, just like every other animal and plant, depend on the natural processes that keep our planet healthy and on the other living things around us. In short, the world leaders had learned the lessons of ecology.

Ecology is the science that explains how plants and animals live together and how they rely on the natural resources around them: sunlight, air, soil, and water. This book is therefore about the processes that make life on Earth possible. It explores the ways in which plants and animals are designed to survive and shows how they all, ultimately, depend on one another. Most importantly, it considers the lessons that we must learn from ecology if the human race is to survive on our small planet orbiting in space.

Exploring the farthest corners of the Earth was a great adventure for the earliest travelers. Exploring how the planet works is an even greater adventure for modern scientists. This book shares in that adventure, and celebrates the amazing variety of plants and animals that are our fellow passengers on Spaceship Earth.

Michael Scott

LIFE-STYLES

All the energy for life comes from the sun.
This energy is trapped by green plants and
convertedinto a form that can be used by other
plants and animals. Grazers eat the plants, hunters
eat the grazers, and recyclers eat their dead
remains, but all face a struggle to survive.

LIVING TOGETHER

Plants and animals do not exist independently of one another. Their lives are closely bound up with other plants and animals, living together in communities to the benefit of them all.

The natural world abounds with examples of the way in which the lives of plants and animals are tied up together, as the case studies opposite demonstrate. Any change that affects one species can have a significant effect on the other plants and animals around it, often in quite unexpected ways.

This idea is best illustrated with an imaginary example. Over a century ago, a scientist named Charles Darwin suggested that there might be a link between the number of cats on a farm and the amount of red clover growing in the surrounding fields, however unlikely that might sound.

Darwin imagined what would happen if all the cats disappeared. Cats are an important hunter of field mice on farms, so, with no cats to hunt them, field mice would quickly build up in numbers. Field mice often attack bees' nests and eat the young bees inside. More field mice would eat more young bees, so there would soon be fewer bees to pollinate the clover. As a result, the clover would produce fewer seeds, and fewer new clover plants would start growing in the fields. Eventually, red clover would become rarer as a result of the cats disappearing.

Living connections

Real farm life is much more complicated than this simple example suggests, but it helps to show the links controlling the lives of all organisms or living things. All the organisms whose lives are connected together in this way make up a community. Thus, in Darwin's example, the cats, field mice, bees, and clover are part of a farmland community. Directly or indirectly, they all depend on each other and on the farmland on which they live.

In the last 20 years, more and more chemicals called pesticides have been sprayed onto rice fields to kill pests that attack the rice. Such large quantities of these chemicals were used that one pest, called the brown planthopper (see inset at right), became resistant and was no longer killed by the pesticides.

The pesticides also killed many of the planthopper's natural enemies, such as spiders and bees. As a result, planthoppers actually became more common and the damage to crops got worse.

In 1986, the Indonesian government banned many pesticides and helped rice farmers to look after their land in a way that encouraged the planthoppers' enemies instead. Spiders, bees, and insect-eating birds increased and planthoppers at last became more rare. By using nature's way, rather than manufactured chemicals, Indonesia is now able to grow 4.5 million extra tons of rice a year.

◁ Cats, mice, bees, and clover are all part of a typical farmland community. But could the cats possibly have any effect on the clover?

NATURE'S WAY WITH RICE

THE UPS AND DOWNS OF AUKS

In the 1970s, the seabirds known as auks, which include razorbills, guillemots, and puffins (see photo), became more common around the North Sea. This increase occurred mostly because the food of these birds – mainly sprats and sand eels – had become more abundant. These small fish had increased because their natural hunters, larger fish such as cod and herring, had been removed by the fishing industry. Now these large fish are running out and humans have started fishing for sprats and sand eels, which are ground up and used as cattle food. Already there are signs that auks are suffering from the over-fishing of these smaller fish. On some Norwegian islands, for example, almost a million puffin chicks starved to death during several summers in the 1980s because their parents could not find enough fish to feed them.

LEOPARDS ARE FRIENDS TOO

Some years ago, people in part of Malawi, Africa, complained that leopards were killing their cattle and dogs. To keep the people happy, the government agreed to kill most of the leopards. Several years later, baboons began to attack the farmers' crops, doing far more damage than the leopards had ever done.

By killing the leopards, people had removed the main hunter of baboons. This allowed the number of baboons to increase, until there were so many that their natural food ran out and they began to attack the farmers' crops. If the farmers had left the leopards alone and made sure that their cattle and dogs were safely protected instead, the leopards would have kept the baboons under control and the "baboon problem" would never have happened.

AS DEAD AS A NUT?

In 1973, a scientist discovered 13 dying trees on the island of Mauritius in the Indian Ocean. He estimated that they were all over 300 years old. They were still producing nuts, but none of these were germinating and growing into new trees.

The scientist remembered that, about 300 years ago, people on Mauritius had killed the last dodo, a turkey-sized bird related to pigeons. He concluded that the nuts would not germinate unless they had first been eaten by a dodo and passed through its gut. Stones in the dodo's crop (a pouch in its gut) may have helped grind up the nuts so that the seedling inside could break out and grow.

The scientist fed nuts to turkeys, which have a similar crop, and found that the seeds in the turkey's droppings grew into new plants. With the dodo extinct, no other bird on Mauritius had been able to crush the hard nuts of the "dodonut tree," so they could not germinate and the tree was in danger of extinction.

Rain forest

Desert

Tundra

Sea (coral reef)

△ ▷ Four of the main ecosystems of the world. Each has its own typical plants and animals, all of which are adapted to live in the special conditions of that ecosystem.

The type of countryside or surroundings in which an organism lives is called its habitat. So farmland is the habitat of field mice in the same way that desert is a camel's habitat and the sea is the habitat of whales. But organisms are also affected by many other factors in the world around them, such as weather, soil, air, water, and the actions of other living things. All these factors together make up the environment in which an organism lives.

Togetherness

The interdependence of plants, animals, and their environment is the basis of the science called ecology. Ecology has been called "the study of the altogetherness of everything" because it helps us understand the rules governing what we call "the balance of nature."

But ecology has another lesson to teach us. In each of the four case studies on the previous pages, it was humans who knocked the system out of balance, and in at least two it was humans who suffered as a result. Ecology shows us that humans are part of the balance of nature and we must learn from its lessons.

This book looks at the natural world in an ecological way. This first section considers some of the factors that con-trol the lives of plants and animals. The second section surveys some of the world's ecosystems: the zones of life in which plants and animals share a similar environment, from the Arctic to the tropical rain forest. The final section investigates the effects that humans have on the natural world, and some of the lessons we need to learn for the future.

▷ The natural world exists in a delicate balance that is easily upset by the actions of humans. The ideas of ecology help us explain how waste gases from a British power station can produce "acid rain" (see page 145) that kills fish in a Scandinavian lake.

ECOLOGISTS AT WORK

Ecologists are scientists who study the relationships between plants, animals, the communities they live in, and the environment around them.

If ecologists were to visit a garden, they would do more than just admire a song thrush singing in a tree. They would want to know what and how much it was eating. They might calculate how much energy it got from this food, and consider whether the food supply limited the number of thrushes living in the garden. They would be interested in what happened to the berries it ate, and whether it was helping more trees to grow by spreading the seeds inside them.

They might investigate whether the number of snails the thrush was eating helped or hindered the growth of those plants normally eaten by the snails. They would look at whether the thrush was competing with squirrels and other birds for food, and how important thrushes were as food for the neighborhood cats.

Eventually, the ecologists would want to understand how the garden community worked. They would want to be able to predict what would happen if all the thrushes were eaten by cats, or all the snails were killed by poison.

▷ These ecologists are analyzing the vegetation growing in a marked-out patch of ground. They might return a year later to see how the plant life has changed over time.

THE NEEDS OF LIFE

Almost all life on Earth depends on energy from the sun. Green plants use sunlight to make their own food, and animals then eat the plants. But plants need more than just sunlight for healthy growth.

If a green plant is moved to a dark place, it will soon become pale and limp. At first, its stem might grow quickly, reaching out for sunlight for its leaves, but soon it will wither and die.

The link between the green color and sunlight is a vital one for the growth of plants and for the survival of almost all living things. The green color comes from a chemical called chlorophyll in the plant's cells. Even plants that do not look green, such as copper beech trees and brown seaweeds, contain chlorophyll, but the green is hidden by other colors.

Powerhouses

The chlorophyll is arranged in even smaller pockets inside the cell, called chloroplasts. These are the chemical "powerhouses" of the plant. They trap the energy of sunlight and turn it into chemical energy. At the same time, the plant takes in carbon dioxide gas from the air, and water from the soil or its surroundings. The trapped energy from the sun is then used to fuel a complex series of chemical reactions. The end product of this remarkable process is sugar, and oxygen is produced as a waste by-product.

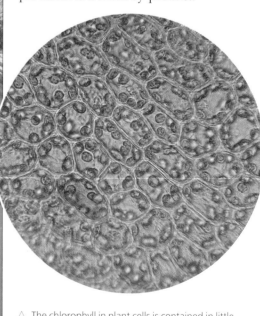

△ The chlorophyll in plant cells is contained in little pockets called chloroplasts – the green blobs inside the cells of this moss.

◁ Even underwater, seaweeds such as this kelp need sunlight to survive. Kelp has a stiff stalk that holds up the leafy blade toward the surface and the sunshine.

This chemical process happens in essentially the same way in all green plants, whether they are single-celled algae floating in a pond or the green leaves at the top of the tallest tree.

The sugar produced is a rich fuel, which the plant can break down quickly to provide the energy for living. Most plants convert the sugar into more complex substances called starches, which store the same amount of energy in less space. The starch can be converted back into sugar whenever the plant needs energy.

This process of using the sun's energy to convert carbon dioxide and water into sugars, starch, and waste oxygen is called photosynthesis. It is not a very efficient chemical reaction. Only about 1 percent of the energy in the sunlight falling on the cell or leaf is turned into stored chemical

▽ Single-celled green algae floating in water sometimes form a green scum called a "bloom" when there are too many chemicals called nitrates in the water. Strictly speaking, algae are not plants but belong to a separate kingdom of life called Protista. However, in this book the word "plant" is used in a general sense for all plantlike organisms whose lives are based mostly on photosynthesis.

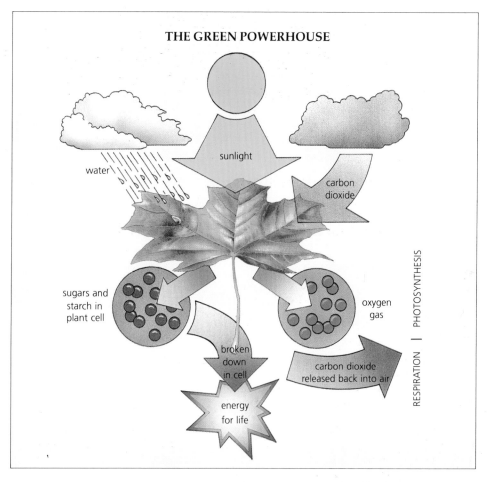

THE GREEN POWERHOUSE

sunlight

water

carbon dioxide

sugars and starch in plant cell

oxygen gas

broken down in cell

carbon dioxide released back into air

energy for life

PHOTOSYNTHESIS

RESPIRATION

▷ ▽ The ghost orchid is a mysterious flower of European beech and oak woods. Its stem is almost transparent, and its leaves are reduced to tiny scales. The whole plant is pink because it contains no chlorophyll and thus cannot photosynthesize. Instead, it relies on a fungus in its swollen roots (see inset below) to break up dead leaves to provide its food.

energy in the form of sugars and starch. But, even so, the world's plants between them produce about 165,000 million tons of sugar a year.

The equally important reverse of this process is called respiration. It is the chemical process by which sugars and other chemicals are broken down by plants and animals to provide energy for living. In the process, carbon dioxide is also released back into the atmosphere. This balance between photosynthesis and respiration powers virtually all life on Earth.

Feeding plant growth

A plant needs more than just sunlight, water, and carbon dioxide to survive. It needs many other chemicals to build the cells and make the many complex chemicals needed for life. Aquatic plants (ones that live in water) can usually get these chemicals from the water around them.

Terrestrial (land-living) plants generally take these chemicals from the soil, using their roots.

Some of the chemicals are required in tiny quantities. For example, the large chlorophyll molecule has one atom of magnesium at its center. Without magnesium, the plant could not make chlorophyll, and thus could not photosynthesize. But it needs magnesium only in small quantities. Too much magnesium in the soil can actually kill plants.

Other chemicals are required in even smaller quantities. They are called "trace elements," because just a trace of them is required. Iron, for instance, is involved in the chemical reactions by which a plant converts starch back into energy. Copper, zinc, manganese, and other trace elements are also required in tiny quantities for healthy growth. If they are missing from the soil, plants may turn yellow, their buds or the edge of their leaves might turn

brown and wither, or the plants might remain short and stunted.

Other chemicals are needed in much larger quantities to help the plant to grow. Nitrogen is particularly important, especially when the leaves and stems are growing, because it is the main component of the building blocks of plant cells – a group of chemicals called proteins. Nitrogen is usually taken in by the plant in the form of nitrates (see page 26). Phosphorus, potassium, sulphur, and calcium are also essential for plant growth, in slightly smaller quantities.

Fortunately, these naturally occurring chemicals, called minerals, are usually plentiful in the soil. However, when a farmer harvests crops from the same field year after year, these minerals are soon used up. The farmer has to replace them, either in the form of natural fertilizers, such as manure, or chemical fertilizers, which are usually made from the three main minerals: nitrogen, phosphorus, and potassium.

▽ In the past, farmers grew a crop of clover in their fields every few years, because clovers and their relatives have bacteria living in nodules in their roots (see inset) that make their own nitrate fertilizers from nitrogen in the air. At the end of the summer, the clover was plowed into the soil, adding nitrates to it and helping crops to grow for the next few years. These days, however, farmers are more likely to spread expensive chemical fertilizers on their fields.

JUST ADD A PINCH OF HYDE PARK...

As long ago as 1699, an English scientist did the first experiment that showed what plants need for healthy growth. John Woodward grew sprigs of a plant called water mint in water he had collected from various places around London. He weighed the plants when he started, and 77 days later he weighed them again to see how they had grown. The drawings below show what he found.

The plants grown in rainwater – the purest source of water Woodward could find – increased their weight by about half, but the ones grown in the muddy water of the River Thames almost doubled their weight. The plants grown in water from a Hyde Park drain did better still, but the best growth of all was shown by the plants grown in the Hyde Park water to which Woodward had added about a table-spoonful of partly rotted leaves from a garden heap. These plants more than quadrupled their weight.

Woodward said that the improved growth was due to increased amounts of "terrestrial matter," but we would now say it was because there were more minerals and trace elements in the water that had drained from land around the Thames or from the rich parkland of Hyde Park. But even the rainwater would have had some minerals in it, from the smoky chimneys of London, and so the plants could grow even in that.

The dead leaves encouraged the best growth of all because tiny plants and animals in the soil – called microbes – were rotting the dead leaves, releasing minerals and trace elements into the water. These helped the mint plant grow strong.

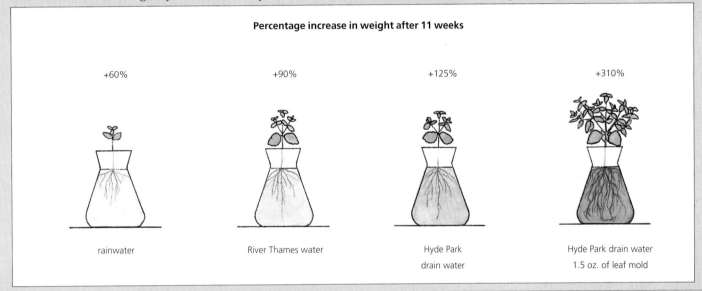

Percentage increase in weight after 11 weeks

+60%	+90%	+125%	+310%
rainwater	River Thames water	Hyde Park drain water	Hyde Park drain water 1.5 oz. of leaf mold

Soil for life

The soil is therefore almost as important to the growth of land plants as is sunlight. It acts like a sponge for the water that plants need for photosynthesis, and it is the main source of the minerals and trace elements they require for healthy growth. It also provides a useful support, helping the plant reach toward the sun so that it can photosynthesize effectively.

Water plants generally get all the chemicals they require from the water around them, but they may still need sand, mud, or rocks to hold them in place. Land plants can also be grown without soil in the laboratory, if they are supported with their roots in water enriched with the right proportions of the essential minerals. Not all soils are equally good for plant growth, however. For example, pure sand could not by itself support any plant growth.

The best soils for plants are those fed by rock that slowly dissolves, releasing minerals into the soil. The gradual rotting and decay of dead plants and animals adds more minerals to the soil, and also adds the fibrous material (called humus) that binds the soil together and helps it hold water.

Rain brings with it more of these essential minerals, carried into the atmosphere

▷ Serpentine rock crumbles to produce a gravelly soil that is poisonous to many plants because of its high metal content. Only a few small plants can cope with these metals. This one – a sort of chickweed – is found only on one patch of serpentine, on an island in the far north of Scotland.

by natural processes such as volcanic eruptions. Over the last century, humans have added to this process. Sulphur, produced by power stations burning coal, comes back to earth, dissolved in rainwater, as a natural fertilizer. Unfortunately, so much sulphur is pumped into the atmosphere that it also has serious harmful effects (see page 145).

The best soils are found over the most crumbly and mineral-rich rocks. The chemistry of the minerals often causes these soils to be slightly basic (the opposite of acidic). Rocks that are harder and less crumbly, such as granite, produce soils that are much poorer in minerals, and these are usually slightly acid. However, they are only a fraction as acidic as vinegar, for example, so sensitive chemical tests are needed to show soil quality. A farmer or gardener who wants to grow crops on acid soil will need to add more fertilizer to the soil for the best results.

In very wet areas, the rain dissolves many of the soil minerals and washes them away, or carries them deep into the ground, beyond the reach of plant roots. The waterlogging of the soil also discourages many of the bacteria, fungi, and soil-living animals that normally break down dead plant and animal matter and release its goodness back into the soil. This makes the soil even more mineral-poor and acid. The result is an acid bog where only those few plants especially good at finding and using the sparse minerals can grow.

At the other extreme, some rocks produce soils that are too rich for plant growth. Soils on a rock called serpentine, for example, are very rich in metals, such as magnesium. Because of these metals, serpentine soils are deadly to most plants. As a result, serpentine rock often produces a bare gravel, with just a few patches of plants able to cope with the poisonous metals.

A RECIPE FOR HEALTHY CRESS

Seeds of cress will germinate (sprout) and grow well on a moist paper towel that has been placed on a dish. They are therefore very useful for experiments to find out what seeds need in order to germinate and grow well.

Make up several dishes and spread about a hundred seeds over each. Enclose each dish in a plastic bag, and make sure the paper inside is kept moist. Place one dish, moistened with tap water, on a windowsill in sunlight. This will be your "control" (standard). You can compare how the seeds on the other dishes grow with this one.

Leave one dish in the refrigerator. Do the seeds germinate? Leave another dish in a dark cupboard. These should germinate, but how do the plants grow? Grow some cress on a piece of flannel moistened only with distilled water (you can buy it in bottles quite inexpensively in supermarkets, pharmacies, and gas stations). To another dish, moistened with tap water, add two drops of house-

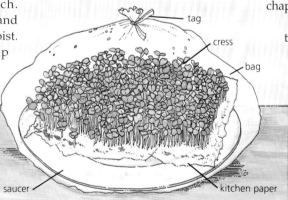

plant food (available from a garden center). Is there any difference in the growth of plants in the distilled water or with added plant food, compared to your "control" on the windowsill? Read the label on the houseplant food to find out what it contains. You have read in this chapter why these substances are needed.

See what other experiments you can think up. For example, shake up a little garden soil with tap water in a jelly jar. Allow it to settle, then add a little of the water on top to your dish of cress. Does this work as well as the plant food?

The nice thing about this experiment is that you can eat your results at the end – but for safety, only eat the ones grown in tap or distilled water.

THE SEARCH FOR NOURISHMENT

Roots must spread far and wide to seek the minerals and other chemicals (called nutrients) that a plant needs for healthy growth. Some plants, however, take shortcuts to find these nutrients.

Many grassland plants, like the yellow lousewort of Alpine meadows, have short roots that penetrate the roots of other plants and steal nutrients to enhance their own growth, without seriously damaging the plant they are attacking. Unlike true parasites (see page 31), these plants make their own food by photosynthesis and steal only a few nutrients, so they are called partial parasites.

Another, less common group of plants, living in poor sandy, rocky, or boggy soils, finds nutrients in a more remarkable way. They are carnivorous – they set traps to catch insects and other small animal prey.

The simplest trap is set by the pitcher plants of American bogs and Asian rain forests. Some of their leaves grow into bottle- or pitcher-shaped traps. Sweet juices produced by the leaves attract insects, which slip on the smooth pitcher walls and slither into a pool of liquid in the base of the pitcher, where they drown. Chemicals in the pool then digest the unfortunate insect, releasing nutrients that the leaf absorbs to help the plant grow.

Sundews also use sweet juices to attract insects to their leaves, which are covered in tendril-like hairs. The insects briefly stick in the syrupy juice, and as they struggle to escape, they brush against the tendrils, which bend over and trap them. Digestive juices then break down the insect to provide nutrients for the sundew.

The most remarkable carnivorous plant is the Venus's-flytrap, which lives in poor, mossy soils in the southern United States. Its leaves are hinged in the middle, and when a suitably sized insect brushes against trigger hairs on their surface, the hinge snaps shut. Digestive juices then dissolve nutrients from the trapped insect.

Yellow lousewort

Pitcher plant

Sundew

Venus's-flytrap

GRAZERS AND HUNTERS

Many animals graze on the abundant growth of plants. They convert the energy that the plants obtain from the sun into flesh. This then provides the food for animals that hunt other animals, in a chain of energy that is a small part of a complex feeding web.

Thanks to the energy of sunlight, plants provide a vast, constantly renewing supply of food for the plant-eating animals of the world. These vary from tiny single-celled animals to elephants and giraffes.

Unfortunately for the plant eaters, however, very little of this food comes ready-to-eat as high-energy sugar or starch. Although these are the first products of photosynthesis, plants convert them into a wide range of other materials, many of which are tough, fibrous, and difficult to digest. Cellulose, the material that makes up cell walls, is the most common ingredient in plants, but it is particularly indigestible for many animals.

The energy stored in starch is so precious that it is usually kept well out of the reach of animals, often in a swollen root or an underground shoot, such as a carrot, turnip, or potato.

▽ A lioness brings down a wildebeest. Lionesses hunt in groups by preying on the weakest animals and killing them after a short chase. The balance of numbers between hunters and grazers is critical if both are to survive.

▷ Many plants offer sweet rewards to animals in return for their unknowing assistance. A hummingbird has been attracted by the sweet nectar of this flower, but as it feeds it may well end up carrying pollen to other flowers, ensuring that they can set seed.

Sugary bribes

Sometimes a plant makes some of its precious sugar available to animals, like a bribe encouraging them to help the plant. Perhaps the most familiar example is the sweet nectar in flowers. This provides a rich supply of energy for hummingbirds and bees, which also use it to make honey. In return, the animals may carry pollen between flowers, helping the plants to set seed and spread.

Many wild fruits are also sweet and sugary. They, too, offer a reward to animals. When a bird, for example, eats a fruit, the seed or seeds inside it pass through the bird's gut and are deposited with its droppings. These then provide a rich fertilizer in which the seeds can germinate.

Plant eaters

Some plant eaters are able to eat the toughest of plant materials, while others eat only the most edible parts. Many of the smaller plant eaters have strong mouthparts, such as bony plates, to grind up the plants they eat. Caterpillars, for example, munch leaves with their strong jaws. Much of this food passes straight through them, but they eat so much that they grow rapidly until they are ready to change into adult butterflies or moths.

Snails and slugs have a ribbonlike tongue that is rough, like sandpaper. They use this to scrape at the leaf surface, breaking off tiny pieces, which are carried into their mouth by the "conveyor belt" of their tongue. These animals are also unusual because they have chemicals in their bodies that can digest cellulose. This enables them to take full advantage of their plant food.

Other small plant eaters feed straight from the plant cells. Some tiny nematode worms, for instance, have a spearlike tube at the tip of their snout, which they use to pierce plant roots. The tube penetrates a cell in the root, and the nematode sucks up

◁ The sugary fruits of plants encourage animals to spread their seeds. Birds like this fieldfare eat the sweet berries of a buckthorn tree and pass the seeds out in their droppings.

the juices in the cell. These juices are much more digestible than the cellulose cell wall. In the same way, aphids suck sap straight from the leaves and stems of plants (see below).

Many of the larger plant eaters, such as humans, have bodies that are adapted to a mixed diet. We can eat fruits and berries, starchy roots like carrots, and the leaves of a few plants like lettuce or cabbage, but we certainly cannot digest grass, one of the most abundant plants.

Most large plant eaters, including humans, have tiny organisms called microbes living in their bodies. Some are bacteria, one of the simplest groups of organisms, some are single-celled animals, and others are microscopic fungi. All of these organisms survive by breaking up the food that passes through the body. They absorb a portion of this food to stay alive, and the rest is made into a form that is easier for their host animal to digest.

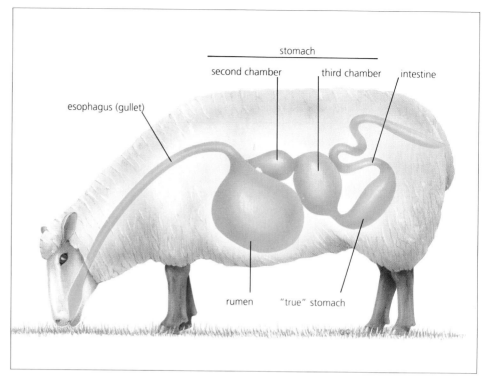

△ When a sheep grazes, the grass is passed to a large chamber at the top end of its stomach, called the rumen, where microbes begin to break down the grass. Later, the sheep "coughs up" the partially digested food (cud) and chews it again, before finally swallowing it. This helps it get the maximum nutrients from the tough leaves.

This, therefore, benefits both the microbes and their hosts.

The grazers

The true grazing animals, called herbivores, are rather well adapted for eating fibrous plant food, although they can be very wasteful. An elephant, for example, digests less than half the plant material it swallows. To stay alive, it needs to eat several hundred pounds of leaves a day.

Only a small portion of the nourishment locked up in all that plant material is released inside the elephant's massive stomach, despite the assistance of its gut microbes. Much of the plant material passes out undigested in the elephant's droppings. But it is not wasted. Dung beetles and other animals feed on the nutrients left in the droppings.

Rabbits have a way of being less wasteful: they eat their food twice! When the rabbit is resting after feeding, it produces soft droppings that have a

A PLAGUE OF APHIDS

Aphids, like these green peach aphids, feed by sucking sap from leaves or shoots with their tubelike mouths. This sap is full of sugars, but it contains very few proteins, the building-block chemicals needed for growth. To get enough protein, aphids suck up far more sap than they need for energy. They pump out the excess from their bodies as syrupy blobs of "honeydew." Cars parked under sycamore or lime trees are soon covered in sticky honeydew, produced by aphids in the trees above.

Although only a fraction of an inch long, aphids can build up in huge numbers. In woods, there can be as many as 5,000 million of them to an acre (equivalent to 30,000 in a piece of woodland the size of this page)!

They can build up their numbers rapidly because female aphids can breed without mating (only a few males are born in late summer). The females give birth to 10 or more live young a day (as the one in the center of the photograph is doing). These daughters can themselves start breeding within 10 days, and if the offspring all survive, a single female can produce 6 million daughters, granddaughters, and so on in two months.

slimy coating. As these are produced, the rabbit quickly bends down and swallows them.

The microbes in the rabbit's body have already broken down the cellulose into a more digestible form before these droppings are produced, so by eating them again the rabbit gets a second chance to absorb the nutrients from its food. The hard, dry droppings that the rabbit finally produces contain all the indigestible leftovers.

Sheep also eat their food twice, but in a different way. Early in the day, sheep busily feed on grass. The eaten grass is stored in a large holding compartment in the stomach, called the rumen. This is full of microbes, which begin to break down the plant food.

Then the sheep coughs up small balls of this partially digested food, called cud, which it grinds and chews with its teeth.

This reduces the food to a paste, which is swallowed again so that true digestion can start.

Animals that chew the cud are called ruminants (after the rumen in their stomach). They include cattle, sheep, goats, antelopes, deer, camels, giraffes, hippopotamuses, and kangaroos. The advantage for these animals is that they can gather food rapidly in areas where they might be in danger from hunters, then retreat to a safe corner to digest their meal.

A chain of energy

In effect, herbivores convert indigestible plant material into a more digestible form of food: their own flesh. Not surprisingly, there are many carnivores (meat eaters) ready to hunt and kill herbivores to get this desirable food.

The result is a chain of feeding. The plants are eaten by herbivores, and the herbivores are eaten by carnivores. But that is not quite the end of the chain: eventually, even the carnivores will die, and they will then be eaten by scavengers.

At the start of the chain is the sun, whose energy is trapped by plants (called primary producers because they produce food from this first source of energy). The plants are then gathered by the herbivores (called primary consumers because they are the first to consume or eat this food). These, in turn, are hunted by carnivores (secondary consumers).

▽ In this food chain, the lettuce makes its own food using the energy of sunlight, the snail feeds on the lettuce, and the thrush eats the snail. However, unless the thrush is wary, it might fall prey to a fox – the top predator in the chain.

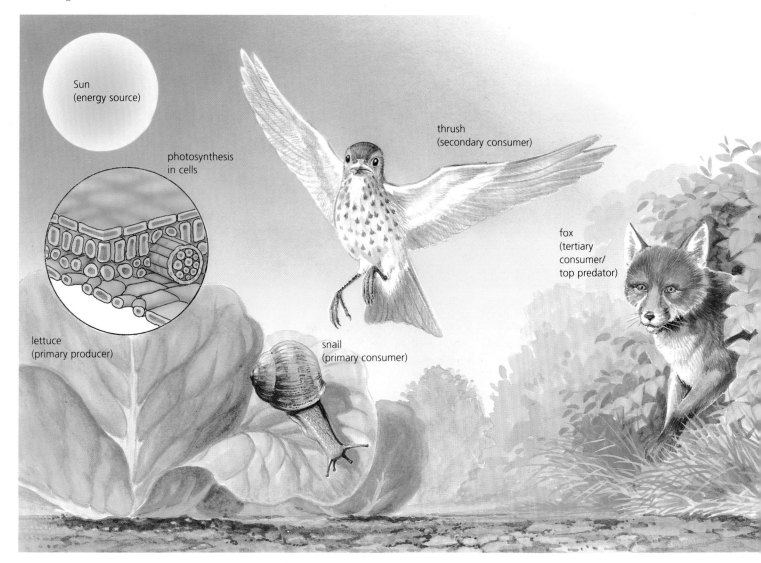

Sun (energy source)

photosynthesis in cells

thrush (secondary consumer)

fox (tertiary consumer/ top predator)

lettuce (primary producer)

snail (primary consumer)

LIFE BEYOND THE SUN

Until very recently, it seemed safe to say that all life on Earth could be traced back to the powering rays of the sun. Sunlight provided the energy for plant photosynthesis, creating the food on which herbivores and, in turn, carnivores fed.

However, in the late 1970s a remarkable exception to this rule was found. Scientists were exploring the seabed 8,200 feet down in the eastern Pacific Ocean in a deep-sea submarine called *Alvin*. What they saw astonished them.

There on the seabed – far beyond the reach of sunlight – they found a bustling community of animals. The powerful spotlights of the submarine picked out "forests" of giant tube-living worms, vast beds of mussels and clams (both species of shellfish), and ghostly gray crabs.

Submarine equipment had already shown that some animals do live on deep seabeds. They feed on the remains of dead plants and animals falling from the surface far above, but this trickle of food is only enough to support a few animals, living far apart on the seabed.

The community that scientists found on the deep-ocean volcanic ridge in the Pacific was very different. There were as many animals there as on the richest coral reef, even though there were no plants in this ever-dark world where photosynthesis was impossible.

So where were all these animals

△ *Alvin* is a deep-sea submarine capable of taking scientists 8,200 feet beneath the ocean.

getting their energy? The scientists concluded that it had to be coming from deep within the Earth. Around these deep underwater ridges huge sections of seabed are moving apart, opening up cracks, or vents, in the Earth's crust. Hot water, heated by the molten rock far below, gushes out of these vents, and provides one source of energy for this strange community.

Bacteria around the mouth of the vents feed on sulphur-rich chemicals that well up through the cracks. The bacteria break these down to release energy, which they use to fuel chemical reactions for making the materials they need to grow.

Other animals graze on the mats of bacteria. These are eaten by carnivores, such as deep-sea fish and octopuses. The crabs and some of the worms around the cracks are scavengers.

Similar communities have now been discovered in other deep-sea vents. Only a few, highly specialized kinds of animals can survive there, because the sulphurous chemicals that pour from the vents would be deadly to most living things.

However, the animals are not the same at every vent. In the Mariana Trough, east of Japan, for instance, hairy snails crowd around the vents, rather than clams and giant tube worms.

The clams have taken their method of feeding one stage further: they give the sulphur-eating bacteria a home within their bodies. The bacteria live in special cells on the clams' gills – curtains of tissue that the clams use to absorb oxygen from the water around them.

◁ Deep-sea vents and their strange animal communities have now been discovered in several places around the oceans; the map shows the main ones in the Pacific.

ASIA

NORTH AMERICA

Juan de Fuca Ridge

Mariana Trough

Pacific Ocean

East Pacific Rise

SOUTH AMERICA

AUSTRALIA

■ main Pacific study areas for vent communities

▷ Worms like this live in tubes of grit on the seabed near some of the deep-sea vents. Sulphur-eating bacteria inside their bodies provide them with food and energy.

△ 8,200 feet down in the Mariana Trough, east of Japan, the spotlights of a submarine reveal hairy snails clustered around the vents, and ghostly crabs scuttling in search of food.

The bacteria provide the clams with energy and nourishment, and in return the bacteria get a safe home. The clams are so dependent on the bacteria providing their food that they have virtually no working gut.

Usually, food chains are quite short. In the African savanna, for example, the vast plains of grass are eaten by antelopes and zebras; they fall prey to hunters such as the lion, and vultures feed on dead lions and the scraps they leave behind.

Antelopes and zebras are also hunted by cheetahs, leopards, and other predators, and lions hunt other herbivores, such as wildebeest and giraffes as well as antelopes and zebras. There are, therefore, a whole series of food chains on the African plains that crisscross together into a complex food web.

Some food chains can be longer than the simple chain on the African plains, especially in the sea, where the primary producers are tiny free-floating plants (see below). However, it is unusual for there to be more than five or six links along any food chain.

Skills of the hunt

To be successful, hunters need to be well designed for their way of life. Above all, they need good senses to find their prey. Generally, they need to be fast, at least over short distances, to catch their prey, unless they find their food by lying in wait and pouncing. In that case, they need a swift strike, great strength, and good camouflage so that their prey will stray

close without suspecting danger. Even fast hunters often benefit from camouflage that disguises them until they are close enough to start a chase.

Flying night hunters, such as owls, have feathers designed for silent flight so that their prey gets no warning of their approach. Some hunters, like lions and wolves, improve their chances of catching their prey by hunting in prides or packs, but they need to catch far more to feed all of the group. Finally, hunters need sharp claws, teeth, or beaks to catch and kill their prey and to tear it apart for eating.

At the top of the pyramid

There are always far fewer carnivores at the top of the chain than there are herbivores farther down the chain. This is easy

▽ Some food chains have many stages. Here, in the southern oceans, single-celled algae floating in the sea build up to large numbers, supported by sunlight and nutrients from the deep ocean. They are eaten by shrimplike animals, which filter water to catch their food. These animals are food for predatory fish, which in turn are preyed on by squid, and the squid might then be eaten by a leopard seal. Nothing hunts the leopard seal, but its dead body might be food for a scavenging skua.

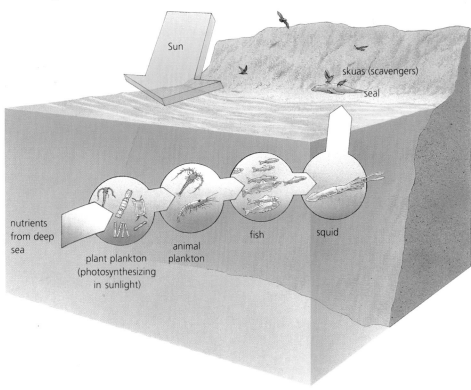

Sun

skuas (scavengers)

seal

nutrients from deep sea

plant plankton (photosynthesizing in sunlight)

animal plankton

fish

squid

to explain if you think of the chain not as a food chain but as a chain of energy. Although the sun provides a vast amount of energy, plants can only convert about 1 percent of this into stored food. Much of the energy that they do obtain is used up just keeping alive, so even less is available to the herbivore that eats the plant.

The herbivore, in turn, will burn off most of the energy it gets from its plant food by moving, breathing, keeping warm, and all the other processes of life, so that a minute portion of the energy that originally came from the sun is left for the top carnivore. The hunting carnivore has an added problem: its food can run away. It therefore needs to expend a great deal of energy just to catch its next meal and restore its energy levels. By the time the carnivore has used its portion to live, the energy chain simply runs out.

This whole process can be illustrated in the form of an energy pyramid. This shows just how quickly energy is used up along a food chain.

The balance between hunters and hunted is a very delicate one. If the hunters catch too many of their prey, their food supply will run out and they will starve to death. They therefore have ways of controlling their numbers to ensure that enough food is always available.

△ Golden eagles usually lay two eggs a few days apart. In good years, both chicks are reared, but in years when food is short, the younger, smaller chick is often killed by its brother or sister. This helps keep the number of golden eagles in balance with their food.

▽ The high food production by plankton in the sea supports much longer food chains in which even predators can be in danger from other predators. Here a killer whale, a top predator, is torpedoing itself up a beach to catch young sea lions, which are hunters of fish that are often themselves predators.

SERENGETI FOOD PYRAMID

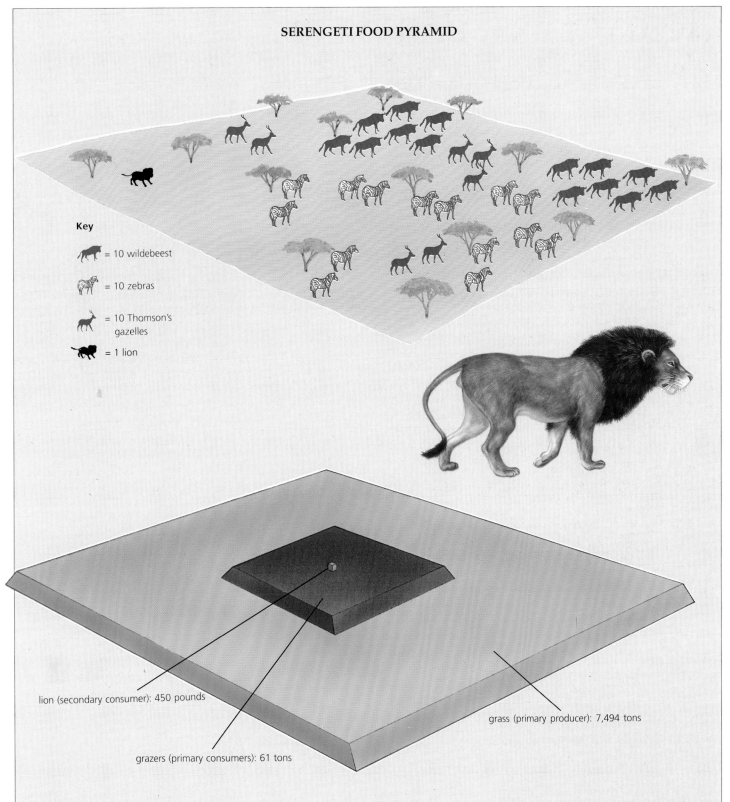

Key

= 10 wildebeest

= 10 zebras

= 10 Thomson's gazelles

= 1 lion

lion (secondary consumer): 450 pounds

grazers (primary consumers): 61 tons

grass (primary producer): 7,494 tons

On the Serengeti Plain of Africa, prides of lions must hunt over a huge area to get enough food. On average, there is one lion to every 5 square miles. The same area holds about 350 wildebeest, zebras, and Thomson's gazelles. With these numbers, hunters and hunted stay in balance from year to year. If there were more lions, they would kill too many of the grazers and run out of food.

If this is made into a diagram showing the weight at each stage of the food chain, the result is called a biomass pyramid (above). A total of 7,494 tons of grass over 5 square miles of the Serengeti support 61 tons of grazing animals, and these provide the food for a 450-pound lion.

RECYCLERS, PARTNERS, AND PICKPOCKETS

Not all plants and animals fit neatly into the simple order of things. Some get their food in devious ways, as nature's "pickpockets." Others have a vital role as nature's "refuse collectors." They recycle dead plants, animals, and their wastes, and thus help the cycle of life to continue.

The chemical process of photosynthesis, which powers life, takes its energy from sunlight (see page 13). In the process, plants use up water and carbon dioxide from the atmosphere. However, both of these are soon returned to the atmosphere by plants and animals, as waste gases from respiration.

Plants need other chemicals called nutrients to nourish their growth. They obtain these nutrients from the soil. Some are re-quired in minute quantity and are soon replaced by the slow breakdown of rocks. Others are required in much greater quantity. The most important of these is nitrogen.

Nitrogen makes up more than three-quarters of the atmosphere, but most plants have no direct way of using this gas.

Instead, they get their nitrogen from the soil in the form of nitrogen-containing chemicals called nitrates, usually dissolved in water. Plants need nitrates in large quantities, and they would soon exhaust all the soil nitrates if these were not replaced.

A cycle of life

There are two natural processes by which nitrates are made. Lightning causes nitrogen to react with oxygen in the atmosphere to create gases, which dissolve in rainwater, forming nitrates. Rain, therefore, brings nitrates into the soil, but only in small quantities.

Some bacteria also have a role in "fixing" nitrogen – taking it from the atmosphere and converting it into nitrates. Some plants, including clover and other members of the pea family, even provide a home for nitrogen-fixing bacteria in nodules on their roots. Thus they have a ready supply of nitrates "on tap."

However, this would not be nearly enough to replace all the nitrates used by plants were it not for the process of re-cycling. As the plant sheds its leaves, and when it finally dies, all the chemicals that

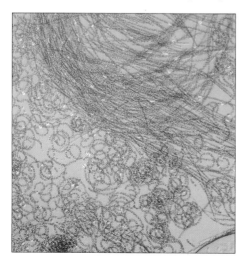

◁ Blue-green algae, a type of bacterium, live in wet places and have an important role in "fixing," or converting, nitrogen from the atmosphere into the nitrates needed for plant growth.

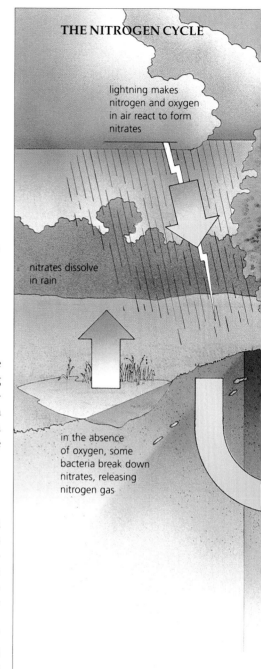

THE NITROGEN CYCLE

lightning makes nitrogen and oxygen in air react to form nitrates

nitrates dissolve in rain

in the absence of oxygen, some bacteria break down nitrates, releasing nitrogen gas

make up the plant body are broken down by microbes and recycled back into the soil as simpler chemicals, such as nitrates. These are then available to support new growth once more. The droppings of animals and the dead animals themselves are also broken down and recycled in the same way. The breakdown of animal and plant matter is called decomposition.

The result is a closed cycle (see above) in which nitrates are used for growth, recycled back into the soil at death, and then become available once more for new growth. All the other main nutrients have

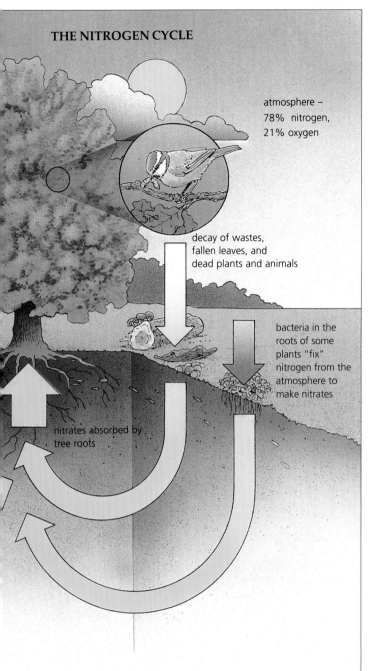

THE NITROGEN CYCLE

atmosphere – 78% nitrogen, 21% oxygen

decay of wastes, fallen leaves, and dead plants and animals

bacteria in the roots of some plants "fix" nitrogen from the atmosphere to make nitrates

nitrates absorbed by tree roots

△ ▽ Flooded paddies produce a large crop of rice, thanks to a tiny fern floating on the surface of the water. The mosquito fern (below) has pockets on its underside that house nitrogen-fixing blue-green algae. The fern is plowed into the soil before the rice is planted, and provides a rich natural fertilizer.

similar cycles, and the keys to this process are the microbes and larger plants and animals that are nature's "waste-disposal experts."

Death and decay

Only about one-tenth of the annual growth of plants is eaten by large herbivores. Nine-tenths of it is eaten by the bacteria, worms, insects, and fungi associated with decay and decomposition.

The process of recycling in nature is similar to what happens at a shipyard, where great ocean liners end their days. A few precious parts of the ship may be taken away and used again. Then the metal will be cut into pieces, sorted out, melted down, and eventually reused in anything from a razor blade to an aircraft wing.

It is the same in nature. "Teams" of decomposers break down the dead plant or animal, bit by bit, until eventually only the simplest chemicals are left in the soil ready to support new plant growth.

When an animal dies, it will first be attacked by scavengers. In Africa, these include hyenas and vultures, while in

Europe and America crows and gulls are more familiar scavengers, feeding at the roadside on animals killed by traffic. They take all the meat that is easily eaten, and perhaps even chew or peck the bones for the nourishing marrow inside.

What they leave is cleaned up by a whole range of smaller decomposers. Bacteria and fungi, in particular, contribute to the rapid decay of the remains. Because these organisms cannot "eat" this solid food by taking chunks of it into their bodies, they release chemicals called enzymes into the body of the animal or the soil around it. These enzymes break down the complex chemicals of the animal body into simple nutrients in a liquid form. The bacterium or fungus then absorbs some of these liquid nutrients as food, while the rest wash away into the soil and are then available for new plant growth.

Plant remains are broken down in the same way, except that there are even more decomposers ready to contribute to this process, especially in woodland areas. The important point is that decay is an essential living process, releasing nutrients and minerals back into the soil to support new life.

◁ The lammergeier or bearded vulture is a mountain scavenger, found in Europe and Asia. It sometimes drops bones from a height of up to 200 feet to smash them open and get at the rich marrow inside.

Line-jumpers

If we imagine a food chain as a line at a supermarket, with each person waiting in turn to get his or her food, then we can also imagine that there might be cheats who would jump the line, or even people who would run off with the food without paying for it at the checkout. There are plants and animals that live in similar ways, but, rather than being cheats, they have an important role in nature.

The "line-jumpers" of the plant world have no green leaves or chlorophyll, so they cannot photosynthesize. Instead, these saprophytes, as they are called, get their nutrition (energy for life and nutrients for growth) from dead and decayed plant matter.

Perhaps the most familiar examples of saprophytes are the larger fungi. The colorful mushrooms and toadstools that we see in woods and grassland are the fungus's equivalent of flowers. They produce spores, which are spread by the wind and eventually grow into new fungi. The real working part of the fungus is the

ROTTING FLIES

Some types of flies, like this blowfly, lay their eggs on dead animals. The larvae (commonly called maggots) hatch within a day or so. They feed by releasing chemicals called digestive enzymes (like the ones that break up food in their stomachs), which turn the tissues of the dead animal into a liquid soup. The larvae grow quickly on this rich food, but they can only eat some of it. A lot washes into the ground, where it can support new life.

After two weeks, the larvae leave the decomposed corpse and crawl underground. There they turn into pupae (the resting stage in which they change into adult insects), before emerging a few days later, ready to fly off in search of a mate and then another rotting corpse in which to lay their eggs.

A blowfly

A rotting wood mouse.

tangle of fine fungal threads that spread out through dead wood or the litter of decaying leaves on a woodland floor and live there even when there are no spore-producing mushrooms or toadstools to be seen above the surface. These feed in the same way as the microscopic fungi and bacteria associated with decay, and help contribute to the recycling of dead plant (or even animal) material.

A few flowering plants – like the ghost orchid on page 14 – also live as saprophytes. Although their relatives have green leaves, they have lost their ability to photosynthesize, and instead rely on decayed plant material for their food. Many live in woods, where the leaf litter provides abundant food. Saprophytes have therefore "jumped the line" by living as decomposers rather than as primary producers.

▷ The fly agaric toadstool, one of the most recognizable of all fungi, is usually found in birch woods because it almost always grows attached to a birch tree. Its fungal threads form a white cloak around the birch roots (inset), gathering nutrients for the tree and protecting the fungus. The red and white cap of the toadstool warns that it is poisonous. This prevents animals from eating it before its precious spores are shed.

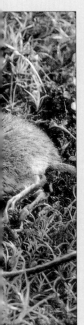

The larvae hatch in a day or two.

The job is all but done and the larvae crawl underground.

Natural partners

Saprophytic flowering plants look strangely similar to woodland fungi. This is no coincidence, since they have a fungus living in their roots and could not survive without it. Flowering plants cannot produce the right enzymes to break down leaf litter, and so saprophytes must rely on nutrients produced by their fungal partner. In return, they offer the fungus the protection of their roots.

Most forest trees also have fungus partners living with their roots, but rather than being inside they form a whitish cloak around the smaller roots. Again, the partnership benefits both the tree and the fungus. The tree gets a ready supply of nutrients, and the fungus probably obtains most of the carbon it needs from sugars moving through the tree roots. Trees could not survive on the poorest soils without their fungal partners, so this partnership is very important in nature, as well as for humans growing tree crops.

Many of the fungi could not exist without their tree partner. Some kinds of toadstools, for instance, are only found in oakwoods, because they can only grow if their mat of fungal threads is attached to the roots of an oak tree.

This system of partnership in nature is called symbiosis. There are many other examples, such as the microbes living in the special "stomach" of cattle and sheep, and the bacteria in the root nodules of clover.

Some partnerships take a different form. Sharks, for example, benefit from "cleaner fish," which follow them around and eat small organisms on their skin. This removes irritation from the shark and provides the cleaner fish with food, to the benefit of both. This, then, is another example of symbiosis.

An even more remarkable example of symbiosis is the way ants "farm" aphids. The ants stroke the body of the aphids, causing them to release honeydew (see page 20), which they then carry back to feed to their young. In return, the ants guard the aphids from predators. Some ants even round up their aphids and take them to the safety of the ants' nest at night.

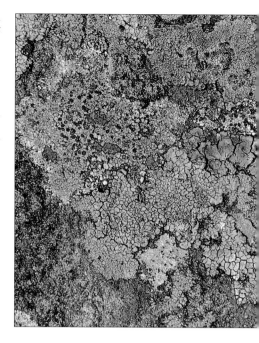

△ Lichens are a remarkable example of symbiosis. Although scientists give them their own species names, they are actually a partnership between a fungus and an alga. The alga lives protected among the fungal threads and photosynthesizes to produce food for itself and the fungus.

▷ Anemone fish live among the tentacles of sea anemones. This sea anemone would normally kill fish this size for food, using the stinging cells on its tentacles, but the anemone fish can shelter itself unharmed among the tentacles, protected by the slime on its scaly skin. The partnership benefits the fish, which can avoid hunters that are not immune to the anemone's stinging cells, and, in return, the anemone benefits from scraps of food left behind by the fish.

◁ Oxpeckers feed on blood-swollen ticks on the hides of African buffalo and other large game animals. The buffaloes benefit from the removal of nasty parasites and get an early warning system from the birds, which call loudly at any sign of danger. The birds get food, a moving perch, and even hair to line their nests!

Pickpockets

Many plants and animals take the process of symbiosis one stage further. They take some or all of the nutrition they require from another plant or animal, but do not offer any benefits in return. These natural "pickpockets" are called parasites, and the plant or animal they attack is referred to as their host.

One familiar parasite is mistletoe, which forms dense green clumps at the tip of tree branches. This plant has no roots connecting it to the ground. Instead, its roots burrow into the branches of its host tree and tap into the supply of water and nutrients rising up the trunk from the roots below.

Mistletoe is only a partial parasite. Its leaves are green with chlorophyll, so it can make its own sugars by photosynthesis. Other plants, such as rafflesia, have no green leaves and must take all their nutrition from their host. They are therefore called total parasites.

Animal parasites are all too common and often unpleasant. Leeches, ticks, fleas, and bedbugs are animal parasites that live by sucking blood, including that of humans. Most animal species have their own kinds of parasites, living on their skin or among their feathers or fur.

To survive, these external (outside) parasites must not kill the host that provides them with food and warmth. They therefore take just a small portion of the food available to them, and are only dangerous if lots of them end up living on the same animal. This usually happens only if the animal is already ill and unable to keep the parasites off.

◁ Rafflesia is a parasitic plant, living on the sap of vines in the jungles of Borneo. It is the world's largest flower, reaching a diameter of 3 feet. Its unpleasant smell, earning it the name "stinking corpse lily," is a trick to attract insects that come to feed on rotting flesh but only end up pollinating the flower.

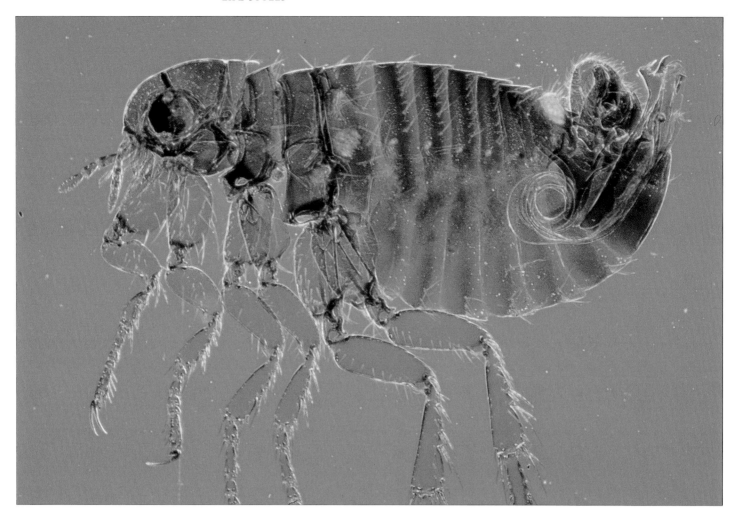

Unfortunately, some parasites can have a more serious and sometimes deadly effect. Female mosquitoes, for example, must suck the blood of mammals before their eggs will fully develop. In the Arctic, vast swarms of these biting flies sometimes drive reindeer wild, but they do them no real harm. For humans, too, mosquito bites are irritating and unpleasant, but not in themselves serious. However, mosquitoes sometimes spread diseases between the creatures they attack. If a female mosquito sucks the blood of a human suffering from malaria, for instance, she may carry the disease to other humans she bites. That is why mosquitoes are considered pests and why in some tropical countries huge sums of money are spent trying to get rid of them.

In fact, the dangerous parasite in this case is not the biting mosquito but the much smaller parasite that it spreads. Malaria is caused by a single-celled organism that lives inside the body of the animals it attacks. It is therefore called an internal parasite. It multiplies rapidly inside the blood cells of its host, until the cells eventually burst. The destruction of these blood cells causes very unpleasant fever symptoms in humans, as well as anemia – a shortage of red blood cells. In severe cases, and especially when young children are attacked, this can lead to death.

Internal parasites like the malaria parasite are often much more dangerous than external parasites, because only part of their life cycle is spent in one host. By the time an animal or a human dies from malaria, for example, mosquitoes will have carried off some of the malaria parasites to other hosts, where they can multiply and spread once more. They therefore have nothing to lose from killing their host.

Some parasites have very complex life histories. The pork tapeworm, for example, lives part of its life in pigs and part in humans who have eaten infected pork that has not been properly cooked. The ribbonlike adult tapeworm attaches itself to the lining of the human intestine

△ The flea is superbly adapted to its parasitic life, with massive piercing jaws and strong legs that enable it to leap between hosts. Its narrow body helps it to move easily through fur.

and simply hangs there, absorbing food passing through the intestine around it. This causes considerable discomfort and weight loss in its unfortunate human host.

Segments of the worm then break off its tail end and pass out in the infected person's feces. These are full of eggs, and if they are then eaten by a pig, the life cycle can begin again. Only good hygiene and careful cooking of pork and ham can control this damaging parasite.

Many other species of tapeworm infect other mammals, birds, and fish. Although the effects of parasites such as these can be unpleasant, they have an important role to play in nature. Like predators, they kill only a relatively small number of their "prey" – often the weakest ones – and thus play a part in maintaining the complex balance of nature.

MYSTERY OF THE LARGE BLUE

In 1979, the large blue butterfly died out from its last breeding sites in England. There seemed to be plenty of the right habitat (rough grassland), with lots of the wild thyme plants on which the butterfly lays its eggs. So why did it disappear?

The answer lies in its amazing life cycle. After the caterpillars hatch, they feed on thyme for about three weeks. Then they drop to the ground and give out a liquid that is attractive to red ants. When a red ant appears, the caterpillar rears up and swells the skin behind its head, tricking the ant into thinking it is one of its own grubs.

The ant carries the caterpillar back to its nest, and it lives in the nest for almost a year, feeding on the ant grubs and spending the winter in the deep sleep called hibernation. In spring, it makes a silk cocoon. While inside the cocoon, it slowly changes into an adult butterfly, before finally leaving the nest in midsummer.

Because grazing by sheep and rabbits had declined on the grasslands where the large blue lived, the vegetation had become too tall and the red ants had been replaced by animals better suited to the new conditions. Although the caterpillars still hatched and fell to the ground, they died, undiscovered by an ant host.

Sheep grazing has now been started up again at these sites, and caterpillars have been brought from Sweden in an attempt to bring back the large blue.

The large blue butterfly

The large blue caterpillar feeds on thyme.

An ant carries the caterpillar to its nest.

The caterpillar lives in the ants' nest.

EAT OR BE EATEN

Plants and animals have a remarkable range of defenses designed to reduce the risk of being eaten. These range from thorns and horns to chemical warfare, and include fast escape and simple trickery.

For a plant, the loss of a leaf or two is not fatal. Indeed, trees can lose all their leaves and still survive to create new leaves (some trees, called deciduous trees, do this naturally every year when they shed their leaves in autumn). However, a great deal of energy and precious nutri-

▽ The leaves of this mullein plant are covered with fine branched hairs. These act like miniature barbed wire and keep small grazers off the leaves.

ents go into making a leaf, so plants have evolved many different defenses to protect their leaves from herbivores.

Thorns and spines are perhaps the most familiar forms of defense, but these are of no use against smaller grazers, such as caterpillars, which just eat around them. A dense mat of hairs or prickles is more effective at keeping caterpillars away. Leathery leaves, like those of holly, or tough stems, like those of rushes, are also effective in discouraging smallgrazers.

Some plants survive by growing hidden among other taller and bushier plants, beyond the reach of ground-living grazers. This has the disadvantage of shading out light, thus reducing their ability to photosynthesize. Many have finely divided or patterned leaves to help hide them among the shadows from grazers.

Chemical warfare

Rather than physical defenses of this sort, many plants rely on chemical warfare! Nettles, for example, have tubelike hairs all over their leaves. When these are touched by an animal, the hairs break and inject a squirt of formic acid into their attacker. This is the same chemical used in ant stings and is more than enough to put off most grazers.

Leaves of many plants contain chemicals that are either unpleasant or poisonous to grazing animals. In some cases, humans actually make use of these chemical defenses. The tea plant, for example, contains two types of chemicals that are unpleasant for animals to eat. However, one of these (tannin) gives tea its refreshing taste as a drink, while the other

(caffeine) is a drug that stimulates us and encourages us to drink more tea.

Similarly, many members of the cabbage family contain unpleasant-tasting oils to discourage grazers, but in the right quantities one of these oils is used to produce mustard to flavor human food. Foxglove is even more valuable. It contains a chemical called digitalis, which is deadly poisonous in quantity, especially to small animals. However, in the right

BRACKEN'S ANT GUARDIANS

Bracken fern is found in every continent except Antarctica, often forming dense patches over hillsides (below). It is a successful weed because it is protected by a range of poisonous chemicals, which discourage most animals from even nibbling it.

Young bracken plants lack these chemical defenses. Instead, they have an animal defense force! The plants give out drops of nectar from small swellings on their stems. These attract ants (see left). To guard their precious nectar supplies, the ants will chase off any grazers, such as slugs or snails, that might damage the plant. Bracken gains valuable protection in return for a little nectar.

stored in separate cells so that they only form cyanide if the leaves are chewed by a grazing animal. Unfortunately, frost can cause plant cells to burst, releasing these chemicals and forming cyanide so that the plant can end up poisoning itself. Plants in frosty northern areas, therefore, do not make cyanide to protect themselves, although luckily there are also fewer grazers in these colder climates.

Because it takes a lot of energy to make these chemicals, many plants produce them only in self-defense. Birch trees, for example, produce chemicals to discourage grazers only when they have been nibbled. Snipping the leaves with scissors will "switch on" this process, but more of the chemicals are produced when caterpillars eat the leaves, perhaps because the plant recognizes the caterpillar's saliva.

STAGGERWORT AND THE STOLEN POISON

Common ragwort, an abundant yellow-flowering weed that grows in grassy fields in many parts of the world (see below), was once called staggerwort because of the effect it had on farm animals. If a horse, cow, or sheep eats a lot of ragwort, it loses its sense of balance and begins to stagger about the field – hence the name staggerwort. By then, however, it is too late to save the animal. Poisonous chemicals in the ragwort will have eaten away the animal's liver almost completely, and death will soon follow.

Most grazing animals, therefore, leave ragwort alone. However, caterpillars of the cinnabar moth (see insets) can eat ragwort without experiencing any ill effects. The poisons from the plant build up inside the caterpillar until it too becomes poisonous, and it stays that way when it turns into an adult moth. Few birds would consider eating a cinnabar caterpillar or moth a second time. The bright orange and black colours of the caterpillars and the red and black of the adults are therefore a warning to leave them alone.

dose, it is used to make the heart work more efficiently. Thousands of human heart patients owe their lives to drugs that are still made from foxglove plants.

Some plants take chemical warfare to extremes. Cyanides are some of the most poisonous chemicals known, and they too are produced in the leaves of plants, such as clovers and trefoils. However, because cyanides are also poisonous to plants, the chemicals from which they are made are

Fleeing and flying

Most animals have the great advantage over plants of being able to run or fly from whatever wants to eat them. Grazing animals get extra protection by feeding in herds. When a predator attacks, they can scatter and run away in different directions, confusing the predator and allowing them to escape.

Being able to move also allows animals to retreat to safety when danger threatens. Many animals have a den in thick vegetation, among stones or underground. They rarely stray far from safety when they feed, and can quickly take cover if danger threatens.

▷ Speed is a great defense. The pronghorn from the deserts and grasslands of the American West holds the long-distance speed record for mammals. It has been timed running at 35 miles per hour over 4 miles, fast enough and far enough to escape hunting coyotes.

Movement also allows animals to defend themselves more actively. Many animals are armed with antlers or horns, which can be used in defense. A healthy reindeer, for example, can often fight off an attacking wolf with its antlers, although if a pack of wolves attacks, the wolves may eventually tire out and kill the deer.

However, horns and antlers are used more often for display fights between males trying to attract a mate than for defense. Females often have smaller horns, but they are just as successful in defending themselves.

Grazers need highly sensitive senses of sight, smell, and hearing so that they can detect and escape trouble. Some animals feed at night, when there are fewer predators around, and then rest in a safe corner during the day.

Hide and seek

Speed of escape is no defense for many smaller animals, whose hunters are longer-legged and faster than they are. They therefore need other advantages to avoid being eaten. One of the most common is for the animals to be camouflaged – that is, colored in such a way that they are hidden against their background.

◁ The leafy sea dragon is a pipefish from the coasts of Australia. It is perfectly camouflaged to escape its enemies. Leafy flaps along its body make it look like a piece of floating seaweed. It even waves about in the water like seaweed drifting in the currents. It feeds on plankton, which is sucked in through its strawlike mouth.

▷ Can you find the bird? The mottled brown color of the woodcock – a relative of curlews and snipe – is perfect camouflage when it nests among dead leaves on a woodland floor.

▽ The greenish color of the three-toed sloth is good camouflage, as it lives in jungle trees. The color is made by algae living in special grooves on its fur.

So, for example, the bright green color of tree frogs camouflages them among the leaves of the bushes in which they live. Many lizards are a speckled brown, which disguises them among stones or against bare earth.

Some animals change their camouflage to suit the season. The brown color of the snowshoe rabbit of North America, for example, camouflages it against rocks in summer, but in winter it grows a new coat of white fur to camouflage it against snow.

Sometimes, only one sex is camouflaged. The males of many species of duck are brightly colored to attract a mate. This might also make them easy for a predator to spot, but they can easily escape onto the water or fly away. Female ducks, on the other hand, are usually a mottled brown color to camouflage them during the dangerous time when they have to sit tight on their nest, with no chance of escaping if a hunting animal spots them.

The shape and behavior of many animals make their camouflage even more effective. A stick insect, for example, is not just the color of a twig but is also shaped like one. This disguise is further enhanced by its shaky way of moving, like a twig blowing in the wind.

△ Shadows can ruin the best camouflage. Many animals, like these impala, have pale, shadowy undersides, which in sunlight look the same color as the rest of their body. This helps to disguise their shape.

When you look at a zebra in a zoo, it is difficult to believe that its black and white stripes could possibly be a camouflage. However, when viewed in the distance on the African plains, the stripes are surprisingly good at breaking up the shape of the zebra's body so that it blends into the heat haze and becomes almost invisible. The stripes may also confuse an attacking lion, which in the rush of the chase may not be able to figure out where one zebra ends and the next one begins – a confusion that just might be enough to allow the zebra to escape.

What works well for the hunted also works for the hunters. Many predators are also camouflaged. Polar bears, for example, are white to disguise themselves against the arctic snow as they creep up on

their prey. Tigers are striped to break up their body shape so they can lurk unseen among the jungle vegetation, ready to pounce on their prey.

A matter of taste

There is still another type of defense for animals that cannot run away or hide. These animals are inedible or poisonous. Many caterpillars, for instance, have a very unpleasant taste. Any bird foolish enough to try to eat one will soon drop it, and it will learn never to attack a similar caterpillar again.

This is taken to an extreme by the poison-dart frog, which produces one of the most deadly chemicals known. Yet the frog is totally harmless – unless you try to eat it! Its poison is there merely to discourage predators. However, by the time a predator has discovered that its prey is unpleasant or poisonous, it may already have taken a bite out of it. To avoid this, many poisonous animals are also brightly colored, which serves as a warning to predators.

Some animals use their chemical weapons in a more active way. Wood ants,

◁ When attacked, the poison-dart frog from the South American rain forest gives out a poison so strong that the tiniest drop would be enough to kill a person. The poison is used only in defense, and the frog's bright "dayglow" colors warn predators to stay away.

INVESTIGATE ANT DEFENDERS

The nests of wood ants are easy to find in most wooded areas where there are pines. The nests are large mounds, often over 3 feet tall, made of conifer needles, leaves, and twigs.

If you approach a nest on a warm day, your footsteps will cause hundreds of ants to swarm onto the surface of the nest to protect it. To find out how they protect it, wave your hand gently back and forth about one inch above the nest, then sniff your hand. It will have the biting smell of the formic acid the ants have sprayed at you. This stinging acid is enough to repel most attackers.

THE SEA SLUG'S STOLEN ARMOR

Sea slugs are among the best defended of all sea creatures, but their armor is not their own. The long fingerlike growths on their backs are armed with stinging cells that come from the sea slug's food.

Sea slugs feed on sea anemones and their relatives. Sea anemones catch their food and protect themselves by using long tentacles packed with masses of stinging cells. Somehow, when a sea slug eats a sea anemone, it stops these cells from firing their stings. The cells then pass along its gut, protected in a coating of slime, and are carried out onto the sea slug's tentacles.

The sea slug is not able to fire its "stolen" stinging cells whenever it wants to, as the sea anemone can. However, if a predatory fish tries to eat the sea slug, it is soon put off by a mouthful of stinging cells.

Sea slug

Sea anemone

for example, spray their attackers with formic acid over a distance of several inches. This acid can eat into the outer case of attacking insects, but more often it just causes pain and discomfort.

The defense of the bombardier beetle is even more effective. When it is attacked, the beetle releases chemicals, produced by its body, into a special funnel-shaped compart-ment near its rear end. A violent chemical reaction takes place, heating up the unpleasant liquid, which explodes out of the funnel and can be squirted directly at its attacker.

Other animals use chemicals for attack as well as defense. Scorpions, for example, use their stinging tails to kill insects, small mammals, and lizards, although the sting is also a useful defense.

◁ Some animals use smell as their defense. A skunk can squirt a foul-smelling fluid over nine feet with amazing accuracy. The spray can temporarily blind its attacker, and will certainly discourage it from threatening a skunk again.

Other animals also make a sacrifice to escape their hunters. Many lizards, for example, shed their tails when attacked. The tails go on wriggling, distracting the hunter while the lizard escapes. It will eventually grow a new tail.

Many other forms of deception are used in self-defense. Some caterpillars, for example, look like snakes, and even rear up and display "fangs" just like those of a real snake. Some toads puff themselves up to almost twice their normal size if attacked and rock from side to side to scare their attacker. Many snakes, including the European grass snake, even fake death to discourage predators, which rarely touch dead animals when they are hunting.

Perhaps the ultimate trickery is shown by plovers. When a predator comes near a plover's nest, one bird outside the nest deliberately shows itself to the hunter, calling loudly and dragging its wing as if it has been injured. The hunter, thinking it has found an easy meal, is lured away from the nest. Once it is well away, the bird "recovers" and flies off, leaving the confused hunter with little hope of finding its way back to the plover's nest.

Only bluffing

The final method of defense in animals is the use of trickery. Some insects, for example, copy the warning colors of poisonous animals, even though these mimics are totally harmless.

Others deliberately set out to shock. A peacock butterfly basks in the sun with its wings partly folded. When attacked, it unfolds its wings, revealing two bright blue eyelike spots. These may fool its hunter briefly into thinking it is attacking a much larger animal, and give the butterfly time to escape.

Some moths have smaller eyespots on their wings, which are there to confuse rather than frighten. A bird may be tempted to peck at the "eyes" and is rewarded with a mouthful of wing, while the moth escapes with only a damaged wing.

◁ The Australian bearded lizard puts on a display to frighten the bravest of predators. It hisses loudly, thrashes its tail backward and forward, and raises the frill behind its head, making it look four times its actual size.

THE EVOLUTION OF A MIMIC

Many animals and some plants protect themselves by mimicking other species that are poisonous, harmful, or unpleasant so that potential enemies will be tricked into leaving them alone. The hornet clearwing moth, for example, looks so much like a stinging hornet that few birds would be likely to attack it. Like the hornet, it even buzzes in flight to complete its disguise.

Mimicry, therefore, clearly benefits the moth, but the intriguing question is how it came to look like a hornet in the first place. The answer lies in evolution, the process by which plants and animals gradually change over thousands or millions of years, and eventually give rise to new species.

Although the young of any animal (or the seedlings of any plant) resemble their parents, they are not identical. Chance twists in their genetic makeup mean that they are slightly different from their parents – and from each other. Sometimes, even a tiny difference might be enough to give one of the young a slightly better chance of surviving to produce its own young. This is what drives evolution.

We can imagine that the ancient ancestor of the hornet clearwing was a typical gray moth. By chance, some may

△ Can you spot the difference? The hornet at the left, one of the largest of the wasps, is feared by many other animals for its unpleasant sting. The hornet clearwing moth at the right is quite harmless, but is such a good imitation of a hornet that birds leave it alone.

have been born with thinner, transparent wings. This would have made them more difficult for hunting birds to spot, so more thin-winged than gray-winged moths would have survived. In turn, some of their young might have had wings that were clearer still, giving them an even better chance of survival. As a result, the gray-winged moths would have gradually been killed off and only the clear-winged ones would have survived.

Another genetic twist might then have produced faint yellow lines like those of a hornet on the bodies of some of the clearwings. These "warning" markings might have made a hunting bird hesitate long enough for the moths to escape, increasing their chances of survival. Over thousands of generations, this process – the "survival of the fittest" – would have ensured that the moths that most closely resembled a hornet would have the best chance of survival, until the modern hornet clearwing moth evolved.

But evolution has not stopped, and hunting by moth-eating birds today is perfecting the mimicry still further.

Because these changes happened in the past, scientists cannot prove that this is how evolution happens, but in the last century another moth has shown evolution at work. The peppered moth of woodlands, parks, and gardens has both a light and a dark form. The light form used to be more common, as its gray mottled markings camouflaged it perfectly against a lichen-covered tree trunk.

However, the development of heavy industry in the 19th century resulted in soot-covered trees, with all their lichens killed. The gray peppered moths were easy for birds to spot on these blackened tree-trunks and were soon eaten. The dark-colored moths, however, were camouflaged, and so far more of them survived. By the 1950s, the gray peppered moth had almost disappeared in sooty industrial areas.

Had industry gone on producing more soot, the gray form might have died out completely. In fact, however, industry in most areas is now much cleaner, trees are recovering their gray, lichen-covered bark, and evolution is once more ensuring that the gray peppered moth is the more common form.

COPING WITH THE COLD

The cold of winter not only brings with it the risk of freezing but also causes many other problems for life. Only highly adapted plants and animals can cope with these chilling extremes.

Ice is an enemy of life. Plant and animal cells are filled with a living fluid called cytoplasm. Cytoplasm can freeze, just like water, although it does so at a lower temperature, about 30.9 °F to 30.2 °F.

One of the remarkable properties of water is that it expands when it freezes.

Because cytoplasm contains a lot of water, it also expands as it freezes. This can cause the cell membrane, the sac containing the cytoplasm within the cell to burst. If many cells are destroyed in this way, a plant or an animal will suffer serious damage or die.

Even before a cell freezes, however, low temperatures cause many of the chemical reactions inside it to slow down or stop. This can kill the plant or animal, unless it is specially adapted to survive.

The cold causes other, more indirect problems. It locks up water as ice, so that animals cannot drink and plants cannot get water for photosynthesis or replace water evaporating into the air from their leaves. Plant photosynthesis is also much reduced because there are only a few hours of very weak winter sunlight. The cold also brings snow, which makes movement and feeding difficult for animals.

▽ Some trees, like these bare birches, survive winter by shedding their leaves. They stay alive by using stored energy in their roots until they can produce new leaves in spring. Many conifers, like the dark green trees seen here, are evergreen. A thick, waxy coating on the leaves and their needlelike shape help reduce the effects of frost.

COLD-BLOODED OR WARM-BLOODED?

Vertebrate animals (those with backbones) are often divided into "warm-blooded" birds and mammals and "cold-blooded" fish, amphibians, and reptiles. In a sense, many invertebrates (animals without backbones) could also be said to be cold-blooded, although the term is not usually used for them, as their blood system is quite different.

Although commonly used, the terms "warm-blooded" and "cold-blooded" are rather confusing, because the body temperature of a cold-blooded animal can sometimes be higher than the body temperature of a warm-blooded animal! The real difference is that birds and mammals are able to keep their body at a steady temperature by producing heat internally. Fish, reptiles, and amphibians, on the other hand, rely on their surroundings to keep their bodies warm, and so their body temperature is much more variable.

The body temperature of most mammals (including humans) remains remarkably steady at 98.4°F, whatever the outside temperature. A variation of more than a degree or so indicates illness. Birds have a higher body temperature of between 100°F to 107°F

To maintain their body temperature, birds and mammals need the insulation of feathers, fur, and fat. However, they inevitably lose some heat in cold weather, and to make up for this they must produce heat. This can come as waste heat from muscle activity – which is why we shiver in the cold – or it is generated by respiring (burning up stored fat).

Cold-blooded animals cannot control their body temperature in the same way, so it varies much more widely according to the temperature of the air or water around them. This is less of a problem for fish, since water is slow to cool or heat up and therefore remains at a much steadier temperature than air.

Reptiles and amphibians on land have to survive much greater changes in temperature, and their scales or skin offer little insulation. In very cold weather, all their body processes slow down almost to a standstill, so that they barely stay alive.

△ The fur and fat of the arctic fox are so effective at stopping heat loss that it can cope with outside temperatures as low as –40°F without increasing the rate at which it burns up fat.

In this state, called diapause, they use up only a little of their fat reserves and can survive for some time.

The timing is critical, however. They cannot stay alive for very long in diapause, because they will soon use up their body fat. That is why there are very few reptiles and amphibians in areas where the winter is long and cold. They need time to adjust to this shut-down state, and a sudden drop to very cold temperatures can be fatal, especially if they have just eaten. At low body temperatures, their digestive enzymes stop working, food rots inside their body, and they die.

When the warmer weather comes, they have to "switch on" quickly. Reptiles will often bask in the morning sun, to warm their body up quickly before going off to feed and restore their lost body fat. Eating in itself helps warm them up, because heat is given out as food is digested. The movement of their muscles helps keep them warm, and reptiles and amphibians also shiver to generate heat.

▽ This adder is basking in the spring sunshine to warm up its body after spending the winter in a state of hibernation.

SLEEPING IT OUT

△ The edible dormouse hibernates through the winter, when berries and nuts are scarce. It feeds up on autumn fruits, especially apples, then rolls up into a tight ball for warmth and "switches off" so completely that it can be picked up without being awakened.

Some mammals spend the winter in a very similar state to the diapause of cold-blooded creatures. Their body processes slow down to one-thirtieth or less of normal levels, their heartbeat drops to as little as 3 or 4 beats per minute (compared with 70 to 80 beats per minute in a human at rest), and they breathe only every two minutes or so.

This process is called hibernation, but, unlike diapause, it remains under the control of the hibernating animal's body. If the outside temperature drops too low, the animal will either wake up or automatically switch on fat-burning to keep its body temperature a few degrees above freezing.

Most hibernating mammals also partly wake up every few days throughout the winter. In springtime, they return to a wide-awake state in response to an internal body clock, even if the outside temperature remains low. To help with these processes, hibernating mammals have a high-energy form of fat, called brown fat, which can be burned up quickly to produce heat.

Small mammals are more likely to hibernate than larger ones, because they tend to lose more heat relative to their body size. As a result, they cannot find enough food to keep alive during long periods of cold weather. Their small size also limits the amount of fat they can store. Mammals that eat insects or invertebrates, such as hedgehogs and bats, are also more likely to hibernate, because their food

△ The whippoorwill, a North American relative of the nightjar, is the only bird known to hibernate. Captive birds can survive for a hundred days in hibernation on just less than an ounce of fat.

supply disappears in winter.

Some mammals, such as squirrels, retreat to a nest in the trees when the weather is really bad, but they do not hibernate. Their body remains active, and they leave their den on milder days to feed.

Because birds can fly, most move away from cold areas in winter, although some birds, such as hummingbirds, can slow their body processes down for a few hours over cold nights without ill effect. However, in 1946, it was discovered that the whippoorwill (a North American bird named after its strange call) hibernates, although only two have ever been found in hibernation because they are so well camouflaged.

Avoiding the cold

Many plants and animals cope with cold simply by avoiding it. They either live in climates where the temperature never drops too low, or they migrate to warmer climates in winter. Those that live permanently in areas with cold winters, however, need special characteristics, or adaptations, that allow them to survive in their environment.

Some animals simply "shut down" for the winter, and many plants survive the winter in a similar state. Some plants (called annuals) die off completely after flowering in late summer, leaving only their seeds to survive through the winter and germinate when the temperature rises in the following spring. To ensure that they do not germinate in warm autumn weather, many seeds must first freeze during the winter (or in a

gardener's refrigerator) before they are ready to germinate.

Other plants, called perennials, continue growing over several years, but they may lose all their leaves or even all their aboveground parts every winter. Life continues only in the roots, safely below the ground, where temperatures never drop very low. Enough energy is stored as starch in the roots for new growth to begin again quickly when spring arrives.

Keeping warm

Some animals have a wide range of adaptations to prevent heat loss and keep their body temperature more or less steady. Even so, most will use up a great deal of stored fat just to stay alive in winter.

A layer of fur or feathers helps to keep precious heat from escaping, by trapping an insulating layer of air. Fluffing up the feathers or fur increases the amount of air trapped and adds to this warming effect.

Many mammals have two layers of hair: soft underhair that traps air and acts as a warming blanket, and longer, coarser outer hairs that protect against the weather. In the same way, penguins are covered by a dense coat of two-layered feathers. At their base, the feathers have tufts of down which trap air warmed by the penguin's body, like an undershirt. At the tip, the feathers are broad and curved, overlapping like tiles on a roof to keep water out.

In cold climates, animal extremities, such as ears and fins, are usually small or well covered, so that little body heat is lost by them. The dense feathers of the snowy

▷ Blubber helps keep these penguins warm when they dive into icy Antarctic waters. Their feathers are densely packed, about 12 to the square inch. Tufts of down at the base of each feather act like thermal underwear, trapping air to keep the bird warm.

owl, for example, cover its legs, toes, claws, and beak. Most animals adapted to cold conditions also have a rounded shape, which is better at preserving heat than a long, narrow shape.

Underwater, fur is less effective for trapping heat. Fat is better at retaining heat, so many sea mammals have a thick layer of oily fat, called blubber, to keep them warm. Their cylindrical shape also helps reduce heat loss. Whales even have what is called a "heat-exchange system," in which their blood is cooled before it reaches their skin, further reducing heat loss.

The final adaptation for many plants and animals is to take advantage of the natural blanket of snow. Because snow traps air around it as it falls, it provides good insulation, and temperatures beneath snow scarcely fall below freezing. Any

◁ The fluffed-up feathers of this pine grosbeak trap an insulating layer of air, which helps keep the bird warm in the cold winter of its home in conifer woods in Canada and the northern United States. During the coldest weather, the grosbeak moves into lower-lying orchards and gardens, where it is warmer.

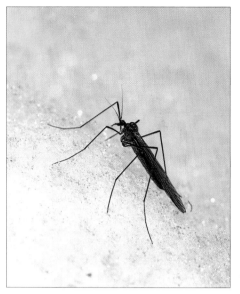

△ Some midges are so well adapted to the cold that they can remain active on snow or glaciers at temperatures as low as 3.2°F.

heat produced by plants or animals is trapped in a warm "igloo" without melting the snow.

Small mammals, like voles, can remain active throughout the winter, feeding on energy-rich plant roots in a network of tunnels under the snow. Even large animals such as polar bears retreat to snow dens in severe weather.

Many plants also survive in cocoons of warmth beneath the snow, ready to burst into growth as the snow melts. Only if the snow is blown off by winds are they at risk from the deadly effects of ice.

△ ▽ The snowbell (above) opens its flowers beneath the snow, so that flowers appear around the edge of a snowpatch as it melts. The dark leaves of the purple saxifrage (below) have absorbed heat from sunlight penetrating the snow. This helps the plant to melt out of a summer snowfall in the mountains and continue flowering undamaged.

DEEP-FREEZE FISH

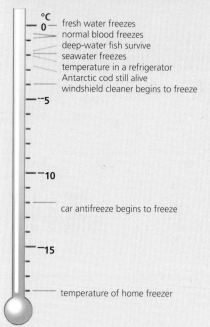

°C
0 — fresh water freezes
— normal blood freezes
— deep-water fish survive
— seawater freezes
— temperature in a refrigerator
— Antarctic cod still alive
— windshield cleaner begins to freeze

5

10

— car antifreeze begins to freeze

15

— temperature of home freezer

Seawater freezes at a lower temperature than pure water because it contains so much salt. The sea temperature near Antarctic ice floes in winter is 28.5 °F, well below the temperature at which blood or cytoplasm should freeze (between 30.9 °F and 30.2 °F).

Marine mammals and birds, such as whales, seals, and penguins, have various adaptations to keep them warm, but why do cold-blooded fish, which cannot control their body temperature, not freeze solid at this temperature?

Part of the answer lies in the way ice forms. Like all crystals, ice needs a nucleus or starting point around which it can build up, like a pearl forming around grit in an oyster. Once an ice crystal begins to form, more water molecules rapidly latch onto it, and an ice mass begins to develop.

△ The blood of the deep-sea Antarctic icefish does not freeze even below 32°F because there are no ice crystals to allow ice to start forming.

Some Antarctic fish avoid freezing by staying deep in the sea. Although the water temperature there can be 28.7 °F, it is a fraction warmer than the freezing point of seawater, and so no ice crystals form in the sea or in the fish. But if the skin or gills of these "super-cooled" fish come into contact with ice, their body fluids begin to freeze immediately and they soon die.

Most Antarctic fish, including the Antarctic "cod," have an even more remarkable system to beat the cold. They have chemicals in their blood that work like antifreeze in a car radiator. These keep their bodies from freezing, even in the lowest temperatures reached by the Antarctic Ocean.

HEAT AND THIRST

Extreme heat can kill plants or animals, but heat is generally less of a problem than the drought that often accompanies it. When it is hot and dry, the main difficulty is saving water.

Death Valley in California is one of the hottest places in the world, with ground temperatures sometimes reaching 188°F. It is also incredibly dry, with just about two inches of rain a year. Yet even here some life survives. Occasional tough, shrubby plants, such as the desert brittle bush and creosote bush, survive the heat and drought, while a few animals rest up in the relative cool of burrows by day, and venture out at night when it is cooler.

The problem for animals in very high temperatures, whether they are warm-blooded or cold-blooded, is that many of the chemical reactions going on in their bodies are temperature-dependent. Very high temperatures destroy the chemical enzymes that control many of these reactions, so that vital life processes stop. Other reactions also speed up in high temperatures, and since these reactions produce heat as well, the animal's body temperature gets totally out of control. One way or another, the result of overheating is death.

Cooling off

Even in less extreme temperatures, most animals need some way of keeping cool. Resting in the shade during the hottest part of the day is often enough. A pale-colored coat also helps by reflecting more heat. Flattening out feathers or fur will also trap less of the animal's body heat.

▽ Death Valley in California is one of the hottest and driest places in the world, yet a few plants and animals survive in this desert hothouse.

◁ In very hot weather, the wood stork from the southern United States lets urine dribble down its legs. The evaporating moisture helps cool it down.

However, in particularly hot weather or after excessive activity, warm-blooded animals in particular may need to lose heat to keep their body temperature steady. The main way to do this is by sweating or panting. As a liquid turns to gas, energy is absorbed and surrounding surfaces are cooled, which is the way refrigerators work. Therefore the evaporation of moisture from the skin in sweating, or from the surface of the mouth and lungs in panting, helps to cool an animal's body.

Some animals take this a step further. Kangaroos, for instance, will lick their front legs, so that heat is lost as the moisture evaporates. Desert tortoises dribble urine over their back legs and saliva over their necks and front legs to cool down.

The other method used to cool the body is similar to the workings of a car radiator. Blood is made to run closer to the surface of the skin so that heat leaks out more effectively. That is why human faces turn red after heavy exercise. Many animals, including African elephants, have large ears for extra surface area for cooling their bodies.

Overheating is less of a problem for birds, partly because their natural body temperature is several degrees higher than that of mammals. Birds able to fly can always take off and let their movement through the air cool them down.

△ ◁ The pale coat of the hairy-footed pygmy gerbil (above) helps stop it from overheating in the desert by reflecting heat from the sun, while its hairy feet minimize contact with the burning-hot sand.
The large ears of the fennec (left) act like a car radiator, losing heat and helping the fox survive in the Sahara Desert.

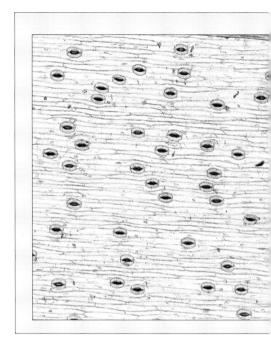

◁ Some seabirds, like this fairy tern from Australia, can slow their heart rate while sitting on the nest as a way of reducing heat production.

Birds on nests, however, pant heavily in hot weather in order to lose heat. They may take as many as 300 breaths a minute (three times their normal breathing rate) to increase the cooling effect. They can spread their wings, which helps shade their bodies and lets more heat escape.

Saving water

In deserts, where water is always in short supply, animals cannot afford to lose water by sweating or panting. At the same time, large ears or other "radiators" are less effective when the temperature is particularly high and the surrounding air is almost as hot as the animal itself.

▽ The African lungfish escapes the hot, dry season embedded in a cocoon in the dried-up mud of its pond. Summer "hibernation" of this sort is called aestivation.

In the absence of any effective way of keeping cool, many desert animals either avoid the heat of the day, or are adapted to cope with a wide range of temperatures without harming themselves. The body temperature of the camel, for example, can rise to 106°F by day and drop to 93°F at night without the camel suffering any ill effects.

In the deserts of North America, a bird called the mourning dove can survive

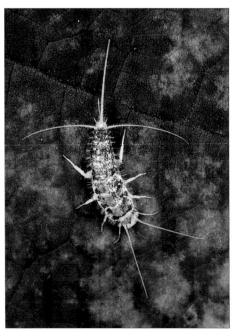

△ Some invertebrates are better at surviving heat than warm-blooded animals. The firebrat, a wingless insect, can live in temperatures of up to 122°F in bakeries, kitchens, and fireplaces.

even if its body temperature rises to 113°F. However, the risk of overheating is always there, and it will die if its body temperature reaches 116°F.

Animals therefore need to save water to help with temperature control as well as for many other needs in their bodies.

Plants and heat

Like animals, plants can suffer from high temperatures, although their life processes produce rather less heat than those of animals, so overheating is less of a threat. The large amount of water in a plant also makes it relatively slow to heat up or cool down. The evaporation of water from the leaf surface in transpiration (see opposite page) helps cool the plant.

The problem for plants is that high temperatures are often combined with drought. To conserve water, plants need to reduce their rate of transpiration by closing the stomatal pores on their leaves, but this also reduces the cooling effect of evaporation and slows photosynthesis.

True desert plants therefore have many adaptations to save and store water, and are able to survive large increases in temperature without damage. Cacti only open their stomata at night, when the air is less hot and dry, and they are able to convert carbon dioxide into a form that can be stored until it is required for photosynthesis in daylight.

WATER OF LIFE

All life on Earth is based on water. Water makes up 70 percent of a typical animal and up to 95 percent of the weight of a plant, and all land plants and animals rely on water for many of the processes of life.

Animals need water to maintain pressure inside their cells. It is one of the main components in the chemical reactions inside their cells, and it is the major constituent of the blood that transports food, oxygen, and waste products around their body. Land

◁ The stomatal pores that regulate the loss of water from a leaf are clearly visible on the surface of this leaf, seen through a microscope.

animals could not breathe were it not for the moisture on their lungs, and many of their waste products are dissolved in water and then passed out of the body as urine.

Plants are just as dependent on water, which is a major component in photosynthesis. Land plants obtain this water from the soil, and with it they pump in all the minerals and nutrients they need for healthy growth. Water is also required to move these chemicals around inside the plant.

The driving force for water intake and movement in plants is a process called transpiration, in which water evaporates from special pores or openings on the

leaf surface, called stomata. The loss of water from the leaf draws water up through the stem, and this in turn allows the roots to take in water from the soil.

The plant also absorbs the carbon dioxide it needs for photosynthesis through these stomata. Therefore, when the plant shuts its stomata to conserve water, carbon dioxide begins to run out and photosynthesis stops. Without transpiration, the roots are no longer able effectively to gather water and minerals from the soil. Land plants are therefore tied to a system that requires them to lose some water if they are to photosynthesize and absorb more water, minerals, and nutrients.

◁ △ The saguaro cactus of the Sonoran Desert is well adapted to save water. It can be 49 feet tall and hold a ton of water. The water cools the inside of the cactus, making a perfect nesting site for this elf owl (above). The hole was originally excavated by a gila woodpecker.

AVOIDING THE EXTREMES

Animals often make long journeys in search of food and shelter. Some animals make regular journeys between two different parts of the world – a process called migration.

△ Two hundred years ago, people believed that, when barn swallows disappeared in the autumn, they went into hibernation in the mud at the bottom of pools. We now know they migrate to southern Africa, where they spend the winter among zebra and wildebeest.

Animals can sometimes overcome the problems of heat and cold by avoiding them. Before the weather gets too difficult, they set off to other places where conditions are better.

Some animals simply wander in search of food. For example, in cold weather European lapwings will fly widely to seek a place where their insect food is not locked up beneath frozen ground. In the same way, short-eared owls will move to lower ground when icy weather in the hills forces the mice and voles they eat to shelter beyond their reach.

Two-way travel

Other animals have a much more regular pattern of movement from one area to another and back again. This repeated, regular, two-way movement is called migration.

Most migration journeys are made between a summer breeding ground and a warmer site to spend the winter. Birds are the true long-distance migrants. The arctic tern, for example, breeds in the Arctic and northern Europe but then travels at least 8,060 miles to spend the southern summer

◁ A locust swarm in New South Wales, Australia. In years when locusts swarm, they travel with the wind to find new areas of vegetation to attack. However, because they do not follow definite routes and most do not return to their home areas, this is not true migration.

WORLD TRAVELERS: SOME ANIMAL MIGRATION ROUTES

Barn swallow

Manx shearwater

Arctic tern

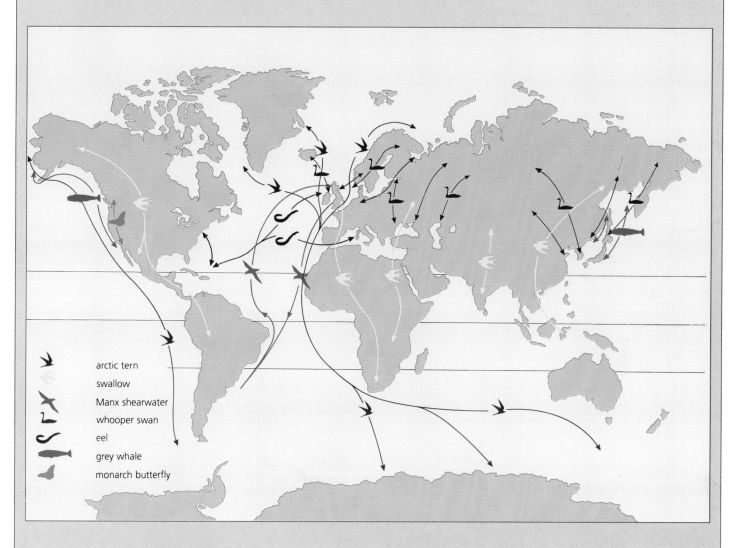

arctic tern
swallow
Manx shearwater
whooper swan
eel
grey whale
monarch butterfly

Whooper swan

European/American eel

Gray whale

Monarch butterfly

near the Antarctic pack ice. The arctic tern therefore lives in perpetual summer.

As seabirds, arctic terns have the advantage of being able to feed on fish as they migrate. Other birds have to travel long distances without feeding. The bristle-thighed curlew, for example, migrates 6,200 miles from its breeding grounds in Alaska to spend the winter on islands in the South Pacific. For at least a third of that journey it is flying over the Pacific Ocean, without any land on which it could stop and feed.

Some migration journeys are much shorter. The American blue grouse spends the winter in mountain pine forests and migrates just 984 feet or so down the hill-sides to nest in lowland deciduous woodland, thus taking advantage of the spring growth of leaves and seeds.

▽ ▷ In May and June each year, half a million wildebeest migrate between the Serengeti Plain of Tanzania, where there is plentiful grass growth during the wet season, and areas farther north, where they can find some grazing even in the long dry season.

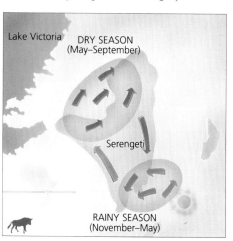

△ ▷ The mass migration of white storks is one of the great spectacles of the animal world. The birds are funneled into narrow crossings (straits) across the Mediterranean Sea at the Bosporus and the Strait of Gibraltar, where up to 12,000 storks have been counted passing in one day.

Other migrants

Birds are by no means the only migrants. In East Africa, wildebeest abandon the grasslands of the plains at the start of the dry season. They trek 149 miles to an area where permanent underground water ensures some grass growth, and where there are some trees to offer shade from the heat.

Whales spend the summer in oceans near the poles, where there is abundant food. Then, before the seas freeze over, they migrate to warmer waters near the equator to spend the winter and produce their calves. Some species of bat also migrate to summer homes where there are plenty of insects to feed both them and their young.

Sea turtles are the longest-distance reptile migrants, traveling between areas of sea that provide rich feeding and coastal areas with safe, sandy beaches where they can lay their eggs.

The Atlantic salmon are the best-known fish migrants. They spend most of their lives at sea, but return to rivers to breed. They may swim more than 992 miles in about six weeks, to the mouth of the river where they were born. They then battle up the river to shallow-water gravel beds, where the females shed their eggs and the males fertilize them. Finally the adults head back downriver. Many die of exhaustion on the way, but those that reach the sea will feed for a year or two before repeating their epic journey to spawn once more.

Some varieties of butterfly migrate. The most famous example is the monarch butterfly, which combines two different ways of escaping the cold. Monarchs breed in the northern United States and in Southern Canada. In July they begin flying south, covering about 80 miles a day, until they reach California, 992 miles away.

FINDING THE WAY

Most animals probably use a variety of cues to find their way on migration. Salmon are guided on their journey by the earth's magnetic fields, sea currents, and the stars, and then "home in" on the distinctive smell (or taste) of the river where they were born. Whales navigate by following magnetic lines around the earth. Satellite tracking of turtles has suggested that they use underwater mountain ranges to guide them.

Most research has been done on bird migration. It has shown that birds mostly use the sun to find their way. This requires that they also have an accurate body clock so that they "know" where the sun should be at any time and judge their direction accordingly. Scientists have shown that if birds are kept in artificial light, it confuses their body clocks and they can be tricked into flying in the wrong direction.

However, when the sun is hidden behind clouds, birds can find their way by using the earth's magnetic field. If a small magnet is attached to a homing pigeon, it will fly in the wrong direction in cloudy weather, because its magnetic navigation system is confused by the magnet. On a sunny day, however, it will find its way correctly using the position of the sun.

Once a bird's navigation systems take it back to the right area, other signals such as smell or characteristic noises may help it find a precise spot, such as its nest site or a pigeon loft.

To find out how night-migrating birds find their way, scientists kept some of these birds in a planetarium, so that they could shift the position of the planetarium stars. The reaction of the birds could then be recorded. The birds were kept in a funnel cage with an ink pad at its base. They therefore left black footprints as they hopped up the side of the funnel, trying to take off. In spring, when the birds should have been heading north, their footprints showed that they spent all their time hopping in the direction of the North Star in the planetarium's sky, proving that they use this star to guide them on their journey.

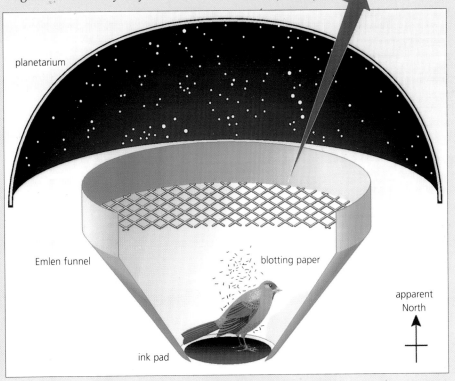

trace on paper inside funnel

planetarium

Emlen funnel

blotting paper

ink pad

apparent North

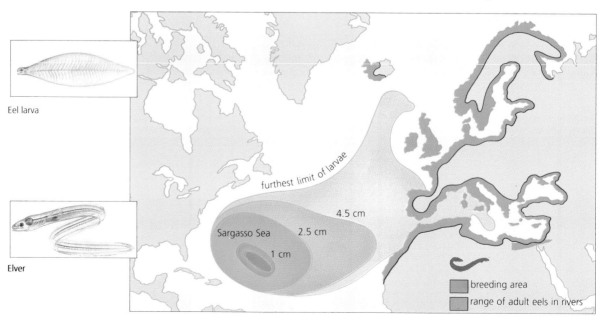

◁ Monarch butterflies spend the winter gathered in their thousands on the trunks and branches of trees in California, having first migrated 992 miles from their breeding sites farther north.

There they settle on tree trunks and branches, often in tens of thousands, and spend the winter in semi-hibernation, becoming active only on the sunniest days.

Why migrate?

It is easy to understand why animals migrate to more favorable areas to avoid drought or winter cold. But why do they ever leave the safety of these areas in the first place to migrate to more inhospitable parts?

In most cases, migrants are seeking greater safety or more food for their young. The brief flush of green plants and the mass hatching of mosquitoes and midges in the arctic spring, for example, provide abundant food for many birds, and this allows them to rear more young. Furthermore, the animals that stay in the migrants' wintering areas throughout the year will also be breeding. There simply would not be enough food to go around if the migrants also stayed there to breed.

The European eel may have a rather different reason for migrating. Eels spend most of their lives in fresh water but return to the sea to breed, crossing the Atlantic to the Sargasso Sea, east of Florida, where they mate and die. The young or larvae of the eels then spend about three years drifting back to the coast of Europe, where they enter rivers to feed and grow until they are old enough themselves to return to the Sargasso Sea. American eels make a similar journey and breed in much the same area of sea, but apparently the young of the two species always head in the right direction to return to their native rivers in America or Europe.

The Sargasso Sea offers a safe hatching place for the eels, with still, deep waters and blankets of seaweed in which the larvae can hide. However, the 3,100-mile journey to breed may be a result of the eel's past. We know that the continents are slowly drifting apart on huge plates that float on the molten core of the earth. Two hundred million years ago, America and Europe were joined as one landmass, but since then they have drifted slowly apart.

When eels first evolved, the Sargasso Sea was much closer to Europe and North America, and their migration was relatively short. Ever since, eels have been returning to the same breeding grounds, even though every year the journey gets a fraction of an inch longer.

▷ Eel larvae look very different from adults. They are leaf-shaped and barely one-half inch long. They drift with the currents back across the Atlantic, growing all the time. This map shows the typical size of larvae as they spread across the ocean. By the time they reach the coasts of Europe or North America, they are about three inches long and ready to change into freshwater-living elvers. These then make their way up rivers to feed and grow.

Eel larva

Elver

furthest limit of larvae

Sargasso Sea

4.5 cm

2.5 cm

1 cm

breeding area

range of adult eels in rivers

TRACKING THE MIGRANTS

Scientists have used many methods to follow migrants. One way is simply to piece together sightings of migrating flocks to find the routes used. Radar screens can help. They show moving flocks of birds, and it is sometimes possible to identify the species from the radar signal.

Animals can be caught and lightweight radio transmitters attached to them so that their movements can be followed from a safe distance. In recent years it has also become possible to attach transmitters that can be followed by orbiting satellites.

The most useful method is also the simplest. For over 80 years, scientists have trapped birds without harming them and fitted lightweight metal bands or rings around their legs. The rings are marked with a number and return address. When a bird is found dead or, more rarely, is caught again, the ring can be returned to the person who fitted it, who can check the original records to find out when and where the bird was tagged.

When thousands of these records are plotted on maps, they give a very good indication of where and when the birds go on migration. They can also provide valuable information about how long birds live, and lead to some unusual record holders. The oldest-known tagged bird is a royal albatross that was fitted with a ring in 1937 and was still alive 54 years later. The long-distance record holder is an arctic tern tagged in July 1955 in northern Russia and found 10 months later 13,968 miles away in Australia.

△ ▷ Birds can be caught in funnel-like traps or in nets called mist nets, which are so fine that they are almost invisible. The birds are carefully picked out of the nets and then a metal ring is gently fitted to their leg with special pliers. The ring carries an identification number and return address. The bird shown being ringed here is a sedge warbler.

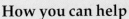

◁ This whooper swan has been fitted with a large plastic ring. The letters on it can be read through binoculars from a distance, so the bird can be identified and its movements followed without having to capture it a second time.

How you can help

To ensure that birds are not hurt, it is illegal for anyone to capture birds for tagging without special training and a license. However, it is just as important to ensure that records of tagged birds are returned. If you find a dead bird, you should check to see if it has a ring on its leg. If it does, carefully take off the ring and return it to the address given on the ring. The person who tagged the bird should always let you know where and when it was first caught.

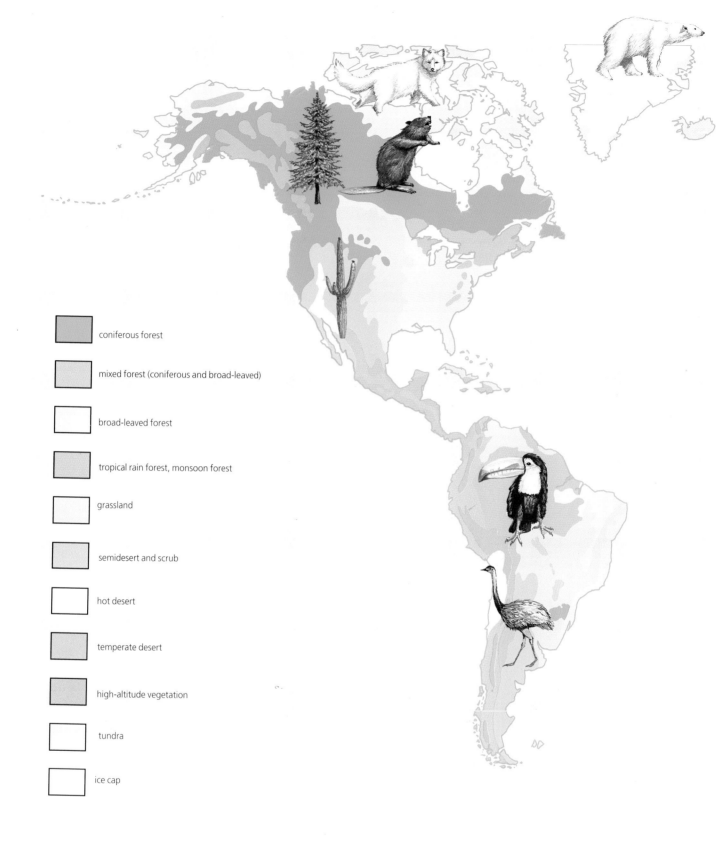

coniferous forest

mixed forest (coniferous and broad-leaved)

broad-leaved forest

tropical rain forest, monsoon forest

grassland

semidesert and scrub

hot desert

temperate desert

high-altitude vegetation

tundra

ice cap

LIFE ON EARTH

From the bitter cold of the Poles to the wet heat
of the tropical rain forest, and from the
mysterious depths of the ocean to windy
mountaintops, the world is divided into a series
of living zones, called ecosystems.
Each zone is the home to a characteristic
range of plants and animals.

LIFE AT THE LIMITS

Life could not be more difficult for the plants
and animals that live near the Poles. It is colder there than
anywhere else on Earth, and ice, snow, and wind
add to the problems they face.

Why do polar bears never eat penguins? The answer to that famous riddle is simple: polar bears only live in the far north and penguins in the far south. But to understand why they live at the opposite ends of the Earth, it is necessary to consider both the similarities and the differences between the North and South Poles.

At both Poles, freezing temperatures during most of the year lock water up as ice, which animals cannot drink and plants cannot use to grow. Snow and ice smother plants, and also make it difficult for animals to travel or search for food.

Fearsome winds batter anything that cannot find shelter, and make the air seem even colder.

For half the year, the sun never rises above the horizon, and the dim daylight lasts for only a few hours. Even in summer, the sun stays low in the sky, giving little warmth, and most of the heat is reflected back into the atmosphere by the shining white icecap.

North and South

Although both polar regions are cold and icy, there are important differences between their weather and landscape, and these differences help to explain why polar bears and penguins live in the regions they do.

Most of the Arctic, including the North Pole itself, is a vast sheet of ice floating on top of the Arctic Ocean. The sea is slow to cool down in winter, so it acts like a warm bath, preventing temperatures from dropping too low. Winter temperatures can reach –88 °F at the North Pole, but that is warm compared with the Antarctic. As a result, the Arctic ice sheet is rarely more than 13 feet thick.

▷ The Antarctic pearlwort (left), a relative of garden pinks, and the Antarctic hairgrass are the only two flowering plants found in Antarctica. Both grow only on the Antarctic Peninsula, a long finger of land surrounded by the warming sea.

The Antarctic, on the other hand, is a huge landmass, rising to over 16,000 feet above sea level. The height and the distance from the sea make it extremely cold: the lowest temperature on Earth was recorded there: –128.5 °F in July 1983. As a result, a huge dome of ice has formed over the continent, so thick that it could bury a mountain. In winter, the sea freezes over and this ice sheet doubles the size of the continent.

Antarctica: sea and ice

Only 2 percent of the Antarctic continent, around the coast and on a few windswept mountaintops, is free from permanent ice. There is no proper soil there, just gravel shattered from the rocks by frost.

Few plants can survive in such an icy, barren land. Only two flowering plants grow in the Antarctic. They live only at the mildest tip of the continent, where they rarely flower, and they were probably taken there by people.

However, 360 species of algae and 400 lichen species are found among the rocks, surviving the winter beneath the snow. Lichens, in particular, grow very slowly in this icy world. Some patches of Antarctic lichens, the size of dinner plates, are over 1,000 years old. The most abundant plants are mosses, which briefly produce green patches among the rocks in summer.

With so few plants to eat, the only grazers are tiny insects and mites, no bigger than pinheads, which feed on dead plant remains. With no large grazing animals, there is nothing for land hunters to eat – which explains why there are no Antarctic polar bears.

With so little living on the land, the sea is vitally important for Antarctic animals. The ocean around the continent is full of life, even though its temperature rarely rises above 32°F. It is very deep, and currents constantly well up from the depths, carrying with them the minerals essential for plant growth.

In this rich water, tiny plants called phytoplankton take advantage of the long hours of summer daylight and build up to huge numbers. They can produce more vegetable matter in an acre of Antarctic ocean than all the plants on an acre of the best farmland in Europe.

Tiny animal plankton feed on the abundant phytoplankton, and they are eaten by larger animals. The most important of

▷ Weddell seals have their pups on ice floes in the Antarctic summer. Fat and slow-moving, they can weigh up to 880 pounds, and are so unused to enemies that scientists can walk right up to them without danger.

◁ The Antarctic is a mountainous continent about one and a half times the size of the United States. Most of the land is buried beneath a vast sheet of ice, which can be up to 2.79 miles thick.

△ Krill are small shrimps that live in the Antarctic oceans in massive shoals, which can turn the sea red for 1,900 feet or more. They are the main food of fish, squid, penguins, seals, and whales.

these are a group of shrimplike creatures called krill. These are eaten by great whales and also by fish, which then become the food of seals and seabirds.

But seals and seabirds cannot breed in the sea. They must come ashore to have their pups or lay their eggs. Thus they become temporarily the main inhabitants of the Antarctic landmass. Directly or indirectly, however, the food for all these animals comes from the sea, a situation quite different from that in the Arctic.

Penguins, therefore, live only in the southern oceans, because these oceans are so much richer in fish than the Arctic Ocean.

Creatures of the ice

Six species of seal breed along the coast of Antarctica, or on ice floes – floating islands broken off the main ice sheet. However, the seals never move far from the sea and always return there to feed.

Seabirds can be seen much farther inland. Adélie penguins, for example, trek over the ice in spring to rocky plains inland, where their nests are least likely to be buried in the snow. Even more remarkably, emperor penguins breed far inland during the height of winter when conditions are at their worst. Gentoo and chinstrap penguins also visit Antarctica in summer, but they are much more common farther north, where the pack ice is less permanent.

A MINIATURE GREEN WORLD

▷ Mossy patches in areas that are free of ice in summer are home for rich communities of tiny Antarctic animals.

▽ Springtails like this one can cope with very low temperatures, thanks to their body's antifreeze. Some have even survived three years, unharmed, inside a glacier.

The few pockets of Antarctica that are free of ice in summer act as miniature oases of life, like water holes in the desert. There melted water allows mosses to flourish, bringing a brief green flush to the bare, rocky landscape. Thirty different species of moss live in the main landmass of Antarctica.

Sheltered among the moss and in cracks in the rocks live minute, blind, wingless insects called springtails, which feed on dead moss. In place of their hind legs, they have a paddle-shaped limb that is folded forward and held by a catch in their hard outer coat. When danger threatens, the catch is released, the paddle flips backward, and the springtail is catapulted to safety.

The biggest problem for the 10 species of Antarctic springtail is the cold. Even in summer, temperatures often drop well below freezing. The mosses in which the springtails live may freeze and thaw as many as 120 times during the summer.

However, freezing is not a major problem for springtails. Like many Antarctic fishes, they have a form of antifreeze in their body fluids, which allows them to survive at temperatures as low as –20 °F without freezing – as long as their gut contains no food or water around which ice can form. Eating is a matter of both life and death for springtails!

◁ The wandering albatross breeds on islands north of the Antarctic. It has a wing-span of up to nine feet.

▽ The emperor penguin is the largest penguin, standing waist-high to a human. Up to 50,000 of them spend the winter huddled together in colonies, incubating their eggs on their feet. This ensures that the slow-growing chicks go to sea for the first time in midsummer, when food is most abundant.

A number of species of gull-like birds feed on penguins. Petrels and fulmars – relatives of albatrosses – only take dead penguins and seals, but south polar skuas are penguin hunters. They wait around the edge of penguin colonies to grab any unguarded eggs or chicks. Skuas are one of the most southerly breeding birds, nesting just 750 miles from the South Pole.

Around the edge of the Antarctic zone, islands such as South Georgia are much milder and support many more species of bird, including the only duck in Antarctica, the South Georgia pintail, and the only songbird, the South Georgia pipit. There, too, are huge breeding colonies of king penguins, a little shorter than emperors, while albatrosses rest and elephant seals breed among clumps of grass.

On northern pack ice

At the other end of the Earth, the permanent pack ice of the Arctic Ocean covers 2,316,781 square miles, expanding in winter to twice that size – as big as the continent of Europe. No plants can grow on the ice, and so the only land animals are wanderers in search of food.

The pack ice is not an unbroken sheet. Winds and sea currents keep a few patches of water open even in winter. These ice-free pools – called polynyas – are vitally important to animals because they allow access to and from the sea beneath the ice.

The Arctic Ocean is almost completely enclosed by land. There are no rich currents flowing into it, like those that feed the Antarctic, and so the seawater is poor in nutrients. Even in summer, the ice shades out 99 percent of the sunlight, limiting the rate of photosynthesis undertaken by phytoplankton. This greatly reduces the food available for animals, and so there is far less life in the Arctic Ocean than in the Antarctic.

Some fish do swim beneath the ice, and a few seals hunt them there. When the seals rest on ice floes or surface in polynyas to breathe, they offer the chance

△ In the worst Arctic snowstorms, polar bears dig dens beneath the snow. Pregnant females disappear into similar dens for up to five months to give birth to their cubs. The cubs do not emerge from their ice den with their mother until they are almost half-grown.

▷ When walruses sunbathe on beaches in summer, veins beneath their skin fill up with blood to keep them cool, causing them to turn pink. When feeding, they dive to 260 feet, plowing up the mud with their tusks to find shellfish.

of food to polar bears. A dead seal is an easier meal, and it is said that a polar bear can sniff one out from a distance of 12 miles.

Mostly, though, the Arctic pack ice is lifeless – it is the world's largest desert.

An arctic feast

Around the edges of the pack ice, however, the Arctic Ocean mixes with the fertile waters of the Atlantic and Pacific oceans. These are rich in plankton, attracting many more fish and their hunters. It is estimated that three-quarters of a million whales, seals, and walruses swim in spring through the Bering Strait – the narrow channel that separates Alaska and Asia – to spend the summer months feasting in Arctic seas.

Even in winter, seabirds like Ross's gulls, Brünnich's guillemots, and little auks fish in polynyas around the edge of the Arctic Ocean. Walruses also gather near polynyas, diving deep beneath the ice to feed on shellfish, which they remove from the mud of the seabed using the bristles on their snouts.

THE UPS AND DOWNS OF LEMMINGS

△ Lemmings are related to guinea pigs, and are about the same size and shape. In winter, they live in runways beneath the snow, eating plant roots, mosses, and lichens. In summer, they also eat grasses, sedges, and tree bark.

In some years, lemmings build up to huge numbers in parts of the Arctic, while in other years they are much less common. These lemming population explosions happen every three to six years. But how do they come about?

In years when the snow forms a deep, insulating blanket and there is plenty of food to eat, female lemmings start breeding in midwinter. By the end of the summer, they will have had as many as

6 litters of up to 13 young, and the youngsters themselves can start breeding after 20 days. This rapidly leads to a huge increase in numbers.

With so many hungry mouths, food becomes scarce, driving many young males to leave home. They wander great distances in search of food, and many die during their travels. There is no truth, however, to the stories of lemmings jumping off cliffs, although they will certainly swim across a lake or river if necessary.

The peak in numbers does not last long. Many predators are attracted to feast on the abundant lemmings. Also, the lemmings spend more of their time fighting than breeding in the overcrowded colony, many young get killed, and others die of starvation. As a result, the number of lemmings drops sharply and takes from three to six years to build up to another peak.

◁ When lemming numbers reach a peak, many predators gather to feast on them. Lemmings can make up 90 percent of the diet of the pomarine skua. They are the main food also for the day-flying snowy owl (left), arctic foxes, and least weasels, which are so slender that they can chase lemmings into their tunnels beneath the snow.

In the Arctic Ocean in spring, a greenish-brown fuzz of microscopic plants, called diatoms, grows on the undersurface of the ice. Many worms, comb jellies, crablike and shrimplike creatures, and young fish feed on these diatoms. These creatures are themselves food for larger fish, which, in turn, are eaten by seabirds, seals, polar bears, and arctic foxes.

With the arrival of spring, cliffs around the Arctic coasts become packed with breeding seabirds, which take advantage of the rich fishing to feed their chicks. Many of the seabirds are related to those of the southern oceans, but in place of penguins of Antarctica there are 14 species of guillemot, razorbill, and puffin – birds collectively called auks.

◁ Comb jellies are egg-shaped, jellylike creatures that move by beating rows of bristles along their sides. These "combs" break up reflected light into a rainbow of colors. They feed on smaller animals in the plankton – including other comb jellies!

▷▽ Fourteen species of auk are found in Arctic waters. Some are extremely common: 4 million pairs of Brünnich's guillemots breed on cliffs and rocky islands, and 17 million pairs of little auks nest among rocks, mostly around Greenland. The great auk became extinct in 1844, as a result of hunting and egg collecting.

Brünnich's guillemot

Great auk

Little auks

△ Despite their woolly coats, musk-ox huddle together for warmth in the coldest arctic weather, scarcely moving to save energy. They are pony-sized beneath their long coats.

▷ Mountain avens hug low to the ground to avoid icy winds in the Arctic summer. The large white flowers follow the sun, like a radio telescope, warming the flowers so that they attract pollinating insects.

Beyond the arctic rim

Encircling the Arctic Ocean is a rim of land. A few animals struggle through the winter there. Musk-ox, for example, have long, woolly coats, lined with underfur that is eight times warmer than sheep's wool. They feed where the vegetation has been stripped bare by the wind.

Reindeer use their hooves and antlers to dig beneath the snow for food. Voles live in a network of runways beneath the snow, searching for roots to eat. Wolf packs hunt both musk-ox and reindeer, and arctic foxes prey on voles, lemmings, and other small mammals.

These land areas are also the true home of the polar bear, which is well adapted to cope with the cold. The transparent white hairs of its coat conduct the warmth of the

sunlight to its skin, which is black for absorbing as much heat as possible.

South of the permanent ice is a region called the tundra, which is milder in summer but still too cold for trees. Even there, the ice is never far away, and 20 feet below the soil surface is a never-melting zone called permafrost.

In summer, the rocky landscape of the tundra is brightened by masses of flowering saxifrages, buttercups, and poppies. Mosses, lichens, sedges, and grasses briefly provide rich feeding for grazing animals. There are also ankle-high shrubs, including a variety of berry-bearing plants. These plants offer a special treat for animals living in the tundra.

A few bumblebees and butterflies emerge to feed on the abundant nectar of the summer flowers, and mosquitoes hatch from lakes and ponds. Now, too, swans, geese, and other birds, which spend the winter in North America or western Europe, briefly move north to the tundra to breed.

But the arctic summer is short. By early September, the summer visitors have left, the snow returns, and the eight-month struggle for survival begins again for the few animals that remain.

Up on the Tops

Plants and animals living high in the mountains have to cope with many of the same problems as those in the Arctic: cold, snow and ice, and fierce winds. As in the Arctic, however, there are many plants and animals well adapted to cope with these problems.

The atmosphere plays an important role for all life on Earth. It acts like a warming blanket, preventing heat from escaping into space. However, most of the air is held close to the Earth's surface by gravity, and so the atmosphere above mountaintops is very thin and its blanketing effects are much reduced.

As a result, temperatures on mountains drop very rapidly, especially at night, and mountaintops can be as cold and windswept as the Arctic. The effect can be seen on Mount Kenya, a huge extinct volcano in Africa. It is situated close to the equator, where the weather at ground level is always hot, but above 15,000 feet on the mountain, there is snow and ice throughout the year.

Cold is not the only problem for life in mountains. The thin air means that there is less oxygen for animals to breathe and less carbon dioxide for plant photosynthesis. At such high altitudes, storms are frequent and there are often strong winds. The thin atmosphere also blocks fewer of the damaging rays from the sun. As a result, human mountaineers can suffer serious sunburn unless they take precautions, and plants and animals need their own "sunblockers" to protect them.

In some ways, therefore, mountains are like a piece of the polar world moved closer to the equator. The two zones even share some of the same plants, which are called arctic alpines. But there is one important difference between the zones. Away from the Poles, there are more hours of daylight in winter and stronger sunshine in summer. Indeed, with such a thin atmosphere to block out the sun, air temperatures can rise quite high in mountains during a summer's day, although they fall quickly at night. Temperatures can therefore vary widely during 24 hours, which is an added problem for mountain plants and animals.

◁ The long, woolly coat of bighorn sheep helps to keep them warm on the cold, windy tops of the Rocky Mountains in winter, but it can be a problem on hot summer days.

Zones of life

All these effects increase with altitude. The average temperature falls by about 34°F every 650 feet up a mountain. Therefore, the higher an animal or plant lives, the more it has to cope with cold temperatures, wind, and damaging rays from the sun.

On the lower slopes, conditions are not very different from those in the surrounding lowlands. The bases of mountains are often covered in dense forests, although in many areas the trees were cut down by humans long ago.

Above a certain altitude, however, the climate is too harsh for tree growth, and the last remaining trees become low and

▷ Mountain glaciers leave behind bare gravel in which only a few arctic-alpine flowers, like this moss campion, can survive.

◁ From the natural woodland in the valleys to the permanent snowfields of the peaks, plants and animals in the Alps are zoned according to their ability to cope with cold and wind, and perhaps with exposure to the harmful ultraviolet rays of the sun.

permanent snow

13,000 feet

lichen zone

upper limit of flowering plants

11,500 feet

upper limit of low-growing shrubs

9,840 feet

upper limit of pasture

8,200 feet

upper limit of coniferous trees

6,560 feet

upper limit of deciduous trees

4,920 feet

3,280 feet

scrubby and eventually disappear completely above what is called the tree line. The altitude at which the tree line occurs depends on the overall climate of the area. In the Alps, for example, the tree line is at about 9,000 feet. In Scotland and Scandinavia, where the climate is windier and the summers are cooler, the tree line may be as low as 1,900 feet.

Above the tree line, there is usually a zone of low-growing shrubs that typically include members of the heather family. Higher still, these give way to the true alpine zone of open grassland, often with a wide variety of flowers. On the highest mountains, even these gradually die out, giving way to a zone of bare rocks and gravel. This is free from ice and snow for such a short time that

nothing can grow, other than a few hardy lichens and algae on the rocks. Above this, the zone of permanent snow and ice begins.

Arctic alpines

Arctic-alpine plants grow near glaciers, both in the mountains and the Arctic. It was glaciers that shaped the landscape of the mountains, and today they still control the plants that grow there. Glaciers leave behind bare gravel in which only a few highly specialized plants can survive. As in the Arctic, the plants must be able to flower and set seed in the brief period when the snow has melted. Many hug close to the ground in cushion shapes to protect them from the chilling and battering wind. They often have small or leathery leaves that conserve precious water, which drains away quickly in this steep and rocky land.

Some arctic alpines contain chemicals that make them unpleasant for grazing animals to eat. Others grow only on cliff ledges, beyond the reach of grazers.

The display of wildflowers on these rugged mountains can be quite beautiful. Saxifrages are the most typical plants of

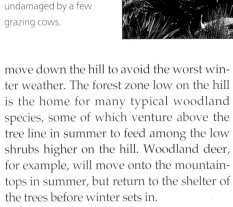

this zone. Their roots grow into rock cracks to anchor them and gather the small amount of nourishment that comes with the rain – the name saxifrage literally means "rock breaker."

Bright-blue gentians, showy yellow- or white-flowered buttercups, low-growing cushion pinks, stonecrops with fleshy water-storing leaves, mountain daisies and dandelions, and tough, spiky sedges all add to the color and variety of the mountain grasslands – and provide some of the most attractive plants grown in rock gardens. Many of these plants live in both the Arctic and the mountains of the Northern Hemisphere, and some are surprisingly widespread.

Near the equator, the mountains of East Africa have a very different flora. Dense rain forests, low on the hills, give way above to thickets of bamboo and then to a grassland zone dominated by giant groundsels and lobelias. These are highly adapted to cope with temperatures that can be boiling hot by day and drop to freezing at night. Only near the snow line does this strange flora give way to open meadows with saxifrages, daisies, and other plants related to the arctic alpines farther north.

△ Meadows high in the mountains support superb displays of colorful flowers, undamaged by a few grazing cows.

Mountain animals

The great advantage that mountain animals have over plants is that they can move down the hill to avoid the worst winter weather. The forest zone low on the hill is the home for many typical woodland species, some of which venture above the tree line in summer to feed among the low shrubs higher on the hill. Woodland deer, for example, will move onto the mountaintops in summer, but return to the shelter of the trees before winter sets in.

The alpine zone itself is home for fewer animals, many of which share some of the adaptations of arctic animals to cope with the cold and wind. These rarely venture down the hill and are only found high in the mountains.

Wild mountain sheep and goats have long, woolly coats to keep them warm, and are amazingly agile as they scramble high onto cliffs in search of plants to eat. Mountain hares and ptarmigan turn white in winter to camouflage themselves against the snow, but they have gray summer coats that hide them among the mountain rocks. The alpine salamander uses its color in a different way. It is a relative of frogs and toads, and like them, it is "cold-blooded."

△ The giant groundsels that grow 13,000 feet up on Mount Kenya are related to the abundant groundsel weeds found in gardens, but they look more like massive cabbages. Their cactuslike leaves spread out in the sun but closeup in the cold of the night. The giant lobelia has similar spreading leaves and tall spikes of blue flowers.

Without fur or feathers to keep it warm, it relies on its black skin to absorb the maximum heat from the sun and it hibernates through the winter. Mountain lizards and snakes also hibernate in winter, and bask on sunny rocks in summer to warm themselves up.

Summer visitors

Although a few larger birds stay on the mountain throughout the year, many others are summer visitors. Songbirds, such as warblers, wheatears, redstarts, and finches, feed on the summer richness of seeds and insects. The ring ouzel of Europe and western Asia is the mountain version of the lowland blackbird, and its

American equivalent is called the mountain bluebird. But these songbirds are only summer visitors; they move to warmer parts in winter.

Among the true mountain birds of the European Alps, the wallcreeper feeds by probing into rock crevices with its thin bill for spiders, millipedes, and other creatures sheltered there. Crag martins – relatives of swallows – build mud nests on rock faces and feed on insects in flight. They migrate to Africa in winter.

A few birds of prey, such as peregrine falcons and some eagles, hunt for birds and small mammals in the mountains. Griffon and bearded vultures in the Alps, and bald eagles and condors in the mountains of the United States, scavenge on dead animals.

With relatively little plant food in the mountains, large grazers like mountain goats and sheep are never abundant, so there are few large hunters. Foxes, lynxes, and wildcats in the European mountains, and bobcats and cougars in the United States, hunt high in the hills, but the supreme mountain carnivore is the snow leopard, a rare big cat of the Himalayas.

A few butterflies and moths take advantage of the summer heat in the mountains. They feed on the nectar of the many flowers that briefly appear, and, like the flowers, they are adapted to complete their life cycle in the short summer months. Grasshoppers and beetles are often abundant in mountain pastures, but only their eggs survive through the winter. Snails also feed on the rich crop of grasses, and hibernate in winter.

▽ The snow leopard has a thick, grayish, spotted coat that helps to camouflage it among the rocks. It hunts mainly wild sheep and goats, following them up and down the mountains as they migrate with the seasons.

▷ The dotterel nests high in the mountains of Europe and in the Arctic. Insect food is so scarce that the female uses up all her stored fat just to produce the eggs. She therefore leaves the male to incubate the eggs and rear the chicks. He is duller-colored than the female, to camouflage him on the nest.

WOODLAND LIFE

Almost two-thirds of the land was once covered with trees, but today more than half of this forest has been cleared away. Many plants and animals still benefit from the shelter and abundant food in the woodland that remains.

In Europe, trees once grew from an area well north of the Arctic Circle almost to the Mediterranean. In Asia and America also, forests spread from the tundra to the parched deserts. This ancient forest was not continuous. There were open areas where the ground was too boggy or the soil too thin and rocky; clearings appeared for a few years after storms blew down trees or lightning caused forest fires; and the higher mountains and windy coasts were free of trees. Even where there was woodland, there were gaps between the trees. The densest forest we find today is mostly the result of changes made by people.

Today, of course, most of this ancient forest is gone. It was cut down for timber and firewood, to make space for farmland,

and sometimes because people thought the forest was a dangerous hiding place for wild beasts.

Most of the remaining woodland has either been planted or greatly changed by people. However, a few precious pieces of woodland still remain, where the trees were not planted by people, the forest still has a wild and natural feel, and a wide range of woodland plants and animals still survives.

Forest zones

Generally, the trees that grow farthest north are conifers. These have needlelike leaves and seeds that grow in woody cones rather than in nuts or berries. (The word "conifer" means cone-bearing.)

Conifers are found as far north as the edge of the Arctic tundra, but the climate prevents them from living any farther north. To grow successfully, conifers need at least 30 days a year when the daily average temperature is above 50 °F. There are no trees in areas colder than this, although there may be a variety of low-growing shrubs (plants that have a woody stem similar to a miniature tree trunk, but are too short to be called as trees).

This northern forest is called taiga. It consists of trees such as pine, spruce, fir, and birch, and it reaches into the far north of Scandinavia and Canada, and to the Taymyr Peninsula in Siberia, 750 miles north of the Arctic Circle.

It is a long way south before conifers are generally replaced by broadleaves as the main forest trees. These are trees with large, flat leaves, such as oak, ash, and beech. They need at least three times as many warm days as conifers to grow well. In Europe, therefore, they mainly grow south of the Baltic Sea.

◁ Although covering barely two square miles, the Bialowieza National Park in Poland is one of the last areas of truly wild woodland in Europe. It is the home of forest animals such as the beaver, lynx, and these European bison.

The large leaf surface of broadleaves means that they "breathe" much more water into the air than narrow-leafed conifers do. In the coldest winter weather, when all the water is locked up as ice, many broadleaves lose their leaves. They survive the winter in a state almost like that of a hibernating animal, storing energy underground in their roots.

Trees that shed their leaves in the autumn are said to be deciduous, but not all broad-leaved trees are deciduous.

Some, such as holly, keep their leaves all year and are called evergreen. Their leaves are thick and leathery to reduce water loss and to protect them from frost. On the other hand, a few conifers, such as larches, are deciduous and shed their leaves.

Climate and soil type are the main factors that determine which tree grows where in broad-leaved forests. Most natural woodland is a mix of many species, with the mixture varying from area to area as the soil and climate change. In Europe, oak or beech are usually the main trees because they are taller and longer-lived than other species. Many other trees and shrubs grow with them and replace them when they die or are blown down, so the pattern of the woodland changes over the years.

▽ Cork is produced from cork-oak trees, grown in open woods in Portugal, Spain, and North Africa. The thick, corky bark, which protects the tree from forest fires, can be stripped from the tree trunk every 10 years or so without damaging the tree.

△ The native woodlands of the Mediterranean have mostly been replaced by a scrub of cistus bushes, as seen here. Cistuses produce richly scented oils that hang in the air above the bushes, like an invisible canopy, helping to reduce the amount of precious water the bushes lose in the dry climate.

Traveling south to the Mediterranean, the forest gradually disappears, often to be replaced by tall bushes with beautiful white or pink flowers and sticky, strongly smelling leaves. These are cistus bushes, and the name given to this type of scrub is maquis (from a French word, pronounced mack-ee). Maquis is a plant community that is probably natural to rocky cliff tops and mountain slopes, but mostly it is a result of people felling the trees since Roman times.

A little natural Mediterranean forest does survive. It is made up mostly of trees such as holm oak and kermes oak. These have evergreen, hollylike leaves, which are leathery to reduce the amount of water the trees lose in the hot, dry summer.

North America and Asia show a similar pattern of zonation to that of Europe, from the northern taiga forest south to the grassy plains and desert. In the United States, broad-leaved forest is found only in the East. In the West, the main trees are conifers, including Sitka spruce and giant redwoods, which can live for 3,000 years.

There is more to a woodland than the trees. Woods are like three-story apartment buildings. Trees make up the top floor, often forming a roof of leaves called the canopy. The middle floor consists of holly, willows, and other shorter trees and shrubs. These cannot outgrow the tall trees but live in the gaps where light penetrates the canopy.

The ground floor of the woodland is called the herb layer, and is made up of flowering plants, ferns, and mosses. The slow-growing ferns and mosses can survive in the dim light that reaches the forest floor in summer. Many of the flowering plants, however, put on a spurt of growth, produce their flowers, and set seed in spring, before the leaves have fully opened in the trees above.

When the leafy canopy is complete in summer, very little light reaches the forest floor, but by then the flowers have passed and their leaves are completely withered. All their energy is stored in bulbs or roots

▷ Bluebells form a colorful carpet in some English woods in springtime, before the leaves have fully opened on the trees. Although abundant in these woods, bluebells are only found on the western edge of Europe, from Spain to England.

◁ Larvae of the buff-tip moth are common caterpillars on oak trees in late summer, sometimes stripping entire branches of their leaves. Altogether, a tall oak tree can support up to 400,000 caterpillars at any one time in summer.

underground, ready to make a quick start to growth when the next spring arrives.

In wetter areas the trunks and branches of the trees are often covered in a carpet of green mosses and their relatives, called liverworts, as well as lichens. Lichens are crisper looking, and grow as a crust, a leafy flap, or a beardlike tuft. They are easily damaged by air pollution, and their healthy growth is a sign of unpolluted air.

Many fungi live in woodlands. Some grow on the trunks of dead or dying trees. Others feed on leaf litter or dead wood, and many have an important role in helping trees to grow by forming mycorhizae with their roots. They bind tightly onto the roots of the tree, forming a close partnership that helps the tree gather nutrients more efficiently from the soil.

Forest feasts

A woodland produces food for animals in each of its layers, from the forest floor to the canopy above. It is also highly productive – in other words, it converts lots of the sun's energy into food. Not surprisingly, therefore, certain animals have evolved to feed on all the different kinds of forest food.

The caterpillars of hundreds of different moths and butterflies feed in huge numbers on leaves. In years when they build up to plague numbers, they can strip all the leaves off trees over large areas. Fortunately, insect-eating birds also quickly increase their numbers in response and soon bring the caterpillars under control, thereby allowing the trees to recover.

On a woodland floor, earthworms, slugs, snails, springtails, mites, ants, fly larvae, and beetles all feed on the rain of dead leaves falling from the branches high above. This breaks the leaves up into tiny pieces, on which other, much smaller animals can feed.

These fragments are then easier for microscopic nematode worms, fungi, and single-celled animals and bacteria to feed on and break down further, in the process releasing nutrients into the soil. There can be as many as 4,000 million bacteria in .04 ounces of woodland soil.

All these animals in the soil are hunted by spiders and centipedes, and when they die they are scavenged by wood lice. The droppings of all these animals help release nutrients back into the soil.

Dead trees are also attacked by decomposers. The fine threads of fungi eat their way through the rotting wood, and their fruiting bodies appear as conspicuous "toadstools" on the side of the tree. The larvae of wood-boring beetles eat their way through the wood, leaving a network of tunnels. Some fungi and beetles also attack living trees and can eventually lead to the death of the trees.

With so many decomposers at work, the nutrients locked up in dead plant and animal material are soon released back into the soil, forming the rich "brown earth" soils that are characteristic of deciduous woodland. But these nutrients are soon absorbed by the trees and other woodland plants, and the cycle of life continues.

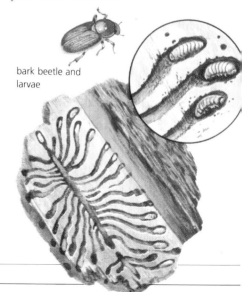

bark beetle and larvae

▷ The larvae of a variety of moths, flies, weevils, and wasps live as leaf miners, leaving their characteristic trails on a leaf once they hatch out.

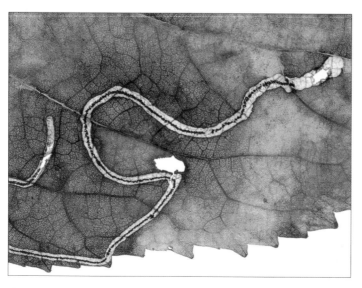

THE DEATH AND LIFE OF A WOODLAND

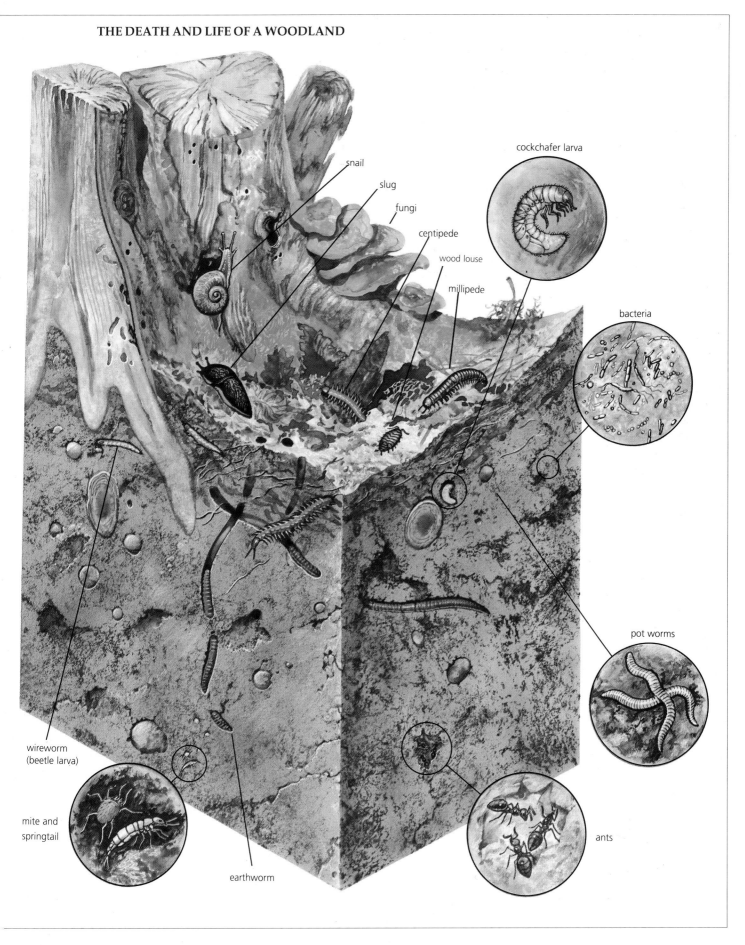

cockchafer larva

snail

slug

fungi

centipede

wood louse

millipede

bacteria

pot worms

wireworm
(beetle larva)

mite and
springtail

earthworm

ants

DISCOVERING THE WOOD'S HIDDEN CREATURES

Many of the most important and numerous woodland animals live hidden from view in the leaf litter on the woodland floor. There are a number of ways to find them.

One is to use what is called a pitfall trap (see right). This is a container hidden in the ground into which small animals moving about the woodland floor might fall. It is important to protect the trap from rain with a raised slate or piece of wood so that the trapped animals do not drown. Also, you should never leave the trap down for more than 6 to 12 hours, or the temporary inhabitants of the trap might starve or eat each other.

When you dig up the trap, empty its contents into a light-colored dish for sorting. Have a close look at them, using a magnifying glass. You may find beetles, centipedes, millipedes, ants, harvestmen (long-legged spiderlike animals), wood lice, and springtails. Make sure at the end to return your captives to the woodland floor, and never leave the trap in place once you have finished with it.

The pitfall trap can only be used to catch animals that move about the woodland floor, not the ones that live buried in the leaf litter. To find these, collect some leaf litter in a plastic bag: scrape aside the top layer of dry leaves, in which few animals live, and collect the damp, rotting leaves beneath.

Take the bag home and empty its contents, bit by bit, into a garden sieve (an old kitchen sieve will do, provided it is not needed again). Shake the smaller animals through the sieve onto a large piece of white paper or an old sheet. Tap the side of the sieve to dislodge animals clinging to the leaves. Finally, turn over the leaves left in the sieve and gently pick out any worms or other creatures too big to pass through it.

You can collect the creatures from the sheet using a sucking tube that you can make yourself with a jar, cork, and rubber tubing. Some of the creatures you may find are shown at the right.

Remember, when you have finished, return all the creatures to the woods from which they came.

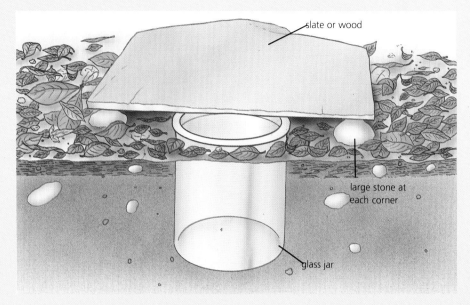

slate or wood

large stone at each corner

glass jar

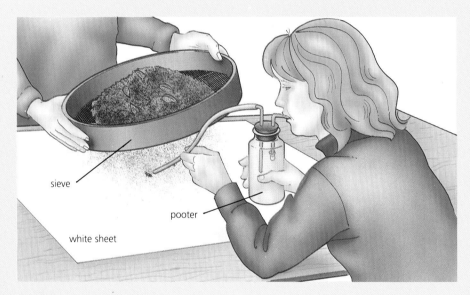

sieve

pooter

white sheet

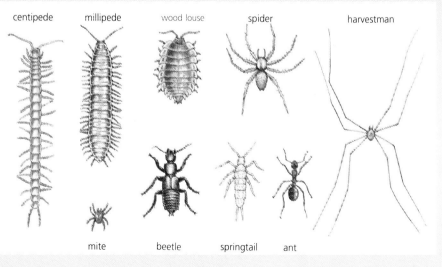

centipede   millipede   wood louse   spider   harvestman

mite   beetle   springtail   ant

▷ Woodpeckers, like this great spotted woodpecker, feed mainly by drilling beneath bark in search of wood-boring beetles. However, some species, such as the ground-living green woodpecker, eat mainly ants.

The larvae of various small moths, wasps, flies, and weevils have a different way of feeding on leaves. They eat into the middle of the leaf, making a narrow tunnel in which they are much safer from hunting birds. These "leaf miners" often have flattened bodies and no legs, so they can squeeze their way into a thin leaf.

Many trees respond to an attack of leaf grazers by producing unpleasant chemicals. However, these chemical defenses cannot be "switched on" until the leaf has opened. It is the young unprotected leaves that are therefore eaten by many insects.

A number of insects, including wasps, midges, weevils, and bugs, can "hijack" the growth mechanism of the plant so that the leaf bud, flower bud or leaf surface swells into a pealike or pill-like growth, called a gall (see also page 80). The gall provides food for the insect's larva and a safe home in which it can grow and eventually turn into an adult. Some eelworms, fungi, and bacteria also make trees produce galls.

Leaf miners and gall insects need secure homes because there are many hunting insects and other small creatures that prey on leaf eaters. Ladybirds are famous as hunters of aphids, but lacewings, beetles, ants, bugs, mites, and bush crickets also feed on leaf eaters. All of these are hunted by insect-eating birds, forming the basis for the complex woodland food web.

The wood and bark of trees are made mostly of lignin, which is hard and indigestible to most animals. However, the strong-jawed larvae of some beetles can munch their way through wood, creating a network of tunnels beneath the bark.

They do little damage to a healthy tree, but can kill old trees.

One group of wood-boring beetles – called ambrosia beetles – "farm" fungi in their tunnels, then eat the fungus's fruiting bodies. The fungus feeds on the beetle droppings, converting the undigested lignin that has passed through the beetle into a form the beetle can eat. The association of beetles and fungi is not always harmless, however. Dutch elm disease (see page 81) is caused by a fungus spread by a beetle.

Woodland birds

With so much food available, the woods are full of birds. Some eat the abundant seeds and fruits. Others feed on the rich supply of insects, while a few hunt the other birds and mammals that take advantage of the woodland feast.

Because insects are in short supply in winter, many woodland birds are summer visitors, such as warblers and flycatchers, which spend the

▷ Crossbills are highly adapted to living in coniferous woodlands. Their twisted beaks are well suited to getting seeds, their main food, from pine, spruce, and larch cones.

winter in Africa or Central and South America. The insect eaters that live in the forest throughout the winter, such as tits, wrens, and tree creepers, find their food by picking over bark or searching under leaves.

Woodpeckers drill into tree trunks with their powerful beaks in search of beetle grubs beneath the bark. The familiar drumming noise of the woodpecker is not made when they are feeding, however. It is a signal warning other woodpeckers to keep out of their territory. The woodpeckers select dead branches that will produce a particularly loud noise, and can drum more than 20 times per second. They have thickened skulls to protect their brain from this violent movement.

Many forest seed eaters are highly specialized for their food. Nutcrackers, for example, are a type of crow with strong, sharp-edged bills. The bills are well suited to cracking open hazelnuts, one of their favorite foods, but are also delicate enough to get the seeds from pine cones.

In autumn, nutcrackers bury extra food as a store to feed their young the next spring. One bird can bury up to 32,000 seeds in its lifetime, but it only retrieves a small part. The rest of the seeds germinate and grow, so the nutcracker has an important role in planting trees. And acorns buried by jays and squirrels are one of the main ways by which oak trees are spread.

Woodland hunters need to be swift and agile. Sparrow hawks and goshawks chase their bird prey through woodland, twisting and weaving among the trees. Owls, which are night hunters, have sharp hearing to find their prey and rely on surprise to catch it. The structure of their feathers makes them almost silent in flight.

Mammals of the forest

There are far fewer mammals than birds in woodlands, although some, like squirrels, are widespread. Squirrels feed on nuts, berries, seeds, catkins, buds, roots, bulbs, and even fungi, while the American gray squirrel also eats the eggs and young of woodland birds.

The forest is the natural home for wild boar and many species of deer. Wolves and bears are the largest forest hunters, while smaller predators include stoatlike pine, beech, and American martens, as well as their relatives, polecats and genets. The pine marten in particular is a fast and nimble hunter, well able to hunt squirrels among the trees.

Perhaps the best-known animal of northern forests is the beaver, which lives along wooded rivers in both Europe and North America. Beavers eat the inner bark, leaves, and twigs of trees such as willow and aspen. They can fell trees with trunks

WHEN APPLES GROW ON OAKS

Oak "apples" are almost as common a sight on oak trees as acorns, which are the oak's own fruit. In fact, oak apples are not fruits but growths called galls. They are produced by the tree when the grubs of a small wasp eat its leaf bud. Up to 30 of the grubs can live safely in an oak apple until midsummer, when they turn into adult wasps and emerge from the gall. The male wasps fly off in search of the flightless females. After mating, the females crawl into the soil beneath the tree to lay their eggs in the tree's rootlets.

△ Oak apples

△ A gall wasp on root galls

When these eggs hatch, a different type of gall develops on the roots. In this gall, the grubs grow slowly over two winters, before turning into wingless female wasps that are able to lay eggs without mating. They climb up the tree trunk to find suitable leaf buds in which to lay their eggs. When these hatch and the grubs start eating the buds, a chemical response causes the tree to produce oak apples, starting the cycle once more.

as thick as a human leg by chiseling around the trunk with their teeth. They use the trees to make dams, and also store them underwater as a food supply when the river is frozen.

◁ The spotted genet is a shy relative of the mongoose, which lives in scrubby woodland in Spain and Africa. It hunts among the branches or undergrowth at night for small mammals, birds, and their eggs.

Dutch elm disease is caused by a fungus that blocks off the tubes carrying sap through the trunk of the elm tree. This starves the growing leaves.The fungus is carried accidentally between trees by a wood-boring beetle that lays its eggs beneath the elm's bark. These hatch into larvae, which eat through the wood, creating a network of tunnels.

The disease has been known since about 1818, but originally it caused only slight damage to the trees. In the 1960s, however, a much more damaging form of the disease appeared in the United States and Canada, where it killed 400,000 trees a year. This was carried to Britain by 1964, probably in imported timber, and it spread rapidly, with deadly effect on British elms. Since 1970, more than 21 million elms have died in Britain, as a result of Dutch elm disease. Lines of dead elms in hedgerows are a common sight.

It is called Dutch elm disease because Dutch scientists were the first to identify its cause. A few strains of elm trees have now been found that are not killed by the disease, but these are still too expensive to plant widely. Otherwise, the only way to stop the disease from spreading is to control the movement of the beetle.

THE ELM KILLER

◁ Alongside their dam, beavers build a lodge of branches and compacted mud that can be as tall as a human, although most of it lies beneath the water. Part of the roof is made only of sticks, to let in air. Entrances underwater lead into tunnels to the central chamber, which is lined with wood shavings. There the two to eight young are born in spring.

SEAS OF GRASS

Where the weather is too dry or the soil too poor for trees, grasses take over. The special way in which grasses grow means they can provide a continuous supply of food for huge numbers of grazing animals.

As any gardener knows, most grasses have shallow roots that spread widely in search of nourishment and water. This helps them to grow in areas where there is too little rain or where the soil is too poor or shallow for trees. In these areas of poor, dry soils, grassland takes over as a natural stage between woodland and desert.

Natural grassland once covered a quarter of the land surface of the Earth. It was known by various names: the steppe in eastern Europe and Asia, the prairie in North America, the pampas in South America, and the savanna in Africa, but in each case grasses were the key to life there.

The mastery of grasses

Grasses have a number of features that make them especially successful. First, their pollen is spread by the wind, not insects, and this helps grasses to produce more seeds than many insect-pollinated flowers. The seeds, too, are often carried by the wind, and this spreads grasses far and wide. To help reduce water loss, many grasses can also roll their leaves up in dry weather to form a moist tube from which little water escapes.

Although grasses are not the most digestible of foods, they have another characteristic that explains why such huge numbers of animals are able to rely on grass as their main source of food. The shoots of many grasses grow less than half an inch above ground level, but, despite being so short, they produce a continuous supply of new leaves. Furthermore, the leaves go on growing from their base even if their tips are eaten.

Because the muzzles of most grazing animals cannot reach very close to the ground, the grass shoots are left untouched as they graze. The grass therefore goes on growing, producing more leaves for the grazers to eat.

This fragile balance between grazers and grass is only broken if grazing becomes too heavy, and grass plants are tugged up by their roots. Once the continuous cover of grass is broken, the soil may be blown away or washed away in sudden downpours.

Natural balances usually prevent overgrazing in nature, but unnaturally high levels of grazing by farm animals have

▷ With wild grazers largely gone, these semiwild gray cattle are the largest grazing animals on the steppes of Hungary.

▷ ▽ There are over 9,000 species of grass, varying widely in appearance and location. Feather grass is a typical plant of the steppes of Europe. Bamboo is a treelike grass that forms dense forests in parts of Asia. Pampas grass, as its name suggests, is a plant of the pampas, and is often grown in attractive clumps in gardens. Wheat, which has been planted in place of wild grasslands in many parts of the world, is an artificial hybrid between two wild grasses.

Feather grass

Bamboo

Pampas grass

Wheat

destroyed grasslands in many parts of the world. Other grasslands have been converted to crop-growing farmland.

Grasslands themselves may just be a passing phase in nature. In some regions, naturally occurring fires are vital to keep the grasslands open. In others, any slight shift in climate can push the grassland either toward desert or woodland. Ancient burrows of grassland animals beneath some woods suggest that trees only began to grow there in the last few thousand years. At the other extreme, huge areas of dry grassland have turned to desert in a series of droughts in recent years. This shift toward desert has been made worse by overgrazing by domestic goats and sheep in these very poor regions.

When the first European settlers reached the American prairies, there were 50 million bison (buffalo) roaming over 525,000 square miles of grassland. Massive hunting reduced their numbers to 1,000 by 1900, but careful conservation has helped increase their numbers to 50,000.

Northern grasslands

In eastern Europe and central Asia, large areas are covered by flat, sandy plains. Lying far inland, without the sea's steadying influence on temperatures, these areas become very cold in winter and very hot in summer. The rainfall is low, and the rain soon drains through the sand. The dryness and winter cold stop trees from growing, producing an open grassland called a steppe.

The wild horses and bison that once grazed these great plains are now replaced by herds of semiwild domestic horses and fierce-looking gray cattle. Europe's only antelope, the saiga, also lives in these dry steppes. Once very rare because

of hunting, it has increased under protection until today there are over a million animals.

Many of the small animals of the steppe live underground in burrows. These include mole rats, marmots, and steppe lemmings. They are hunted by hawks, owls, and polecats.

Steppe birds include the great bustard. These turkey-sized birds, with a wingspan of up to 8.5 feet, once gathered in huge flocks on the steppes. Today they are much rarer,

partly because of hunting but mostly because of changes in their habitat. Modern farming destroys their breeding sites, while many of these clumsy, low-flying birds are killed by power lines or on roads. Even hedges and planted trees can be a problem; bustards need more than half a mile of open view in at least three directions before they will settle.

The American equivalent of the steppe is the prairie, which once occupied a huge area in the center of the country. Most of the prairie today has been plowed up to grow wheat, making this part of the

The male great bustard performs an extraordinary mating display to attract females on the central European steppes. He turns his tail back to display the white feathers underneath and puffs up his feathery throat so that his head sinks into a shimmering white powderpuff.

The saiga antelope's long nose may help to warm the air it breathes in the cold winter on the steppes, and moisten the air during dry summer weather. It can also close its nostrils during dust storms.

United States one of the great "breadbaskets" of the world.

Some natural prairie does survive, and many of the animals that live there look strikingly similar to those of the steppes. In place of the saiga is the antelopelike pronghorn (see page 36), while ground squirrels and prairie dogs (see below) are the American equivalent of the marmots and lemmings of the steppes. Huge herds of American buffaloes once roamed the prairies, and were hunted by wolves and coyotes, although both the bison and their hunters are now rare.

The prairie was once alive with colorful flowers, with descriptive names like prairie cat's-foot, prairie lily, Indian paintbrush, and black-eyed Susan. Huge numbers of insects visited and pollinated this sea of flowers. Today, most of that prairie color has gone, like the prairie itself, although in some areas scientists are trying to recreate patches of flower-rich prairie grassland.

THE BANDIT IN PRAIRIE DOG CITY

Prairie dog

Prairie dogs are named after their bark or warning call, although in fact they are ground-living squirrels. They live in huge colonies, each of which occupies a maze of burrows beneath the prairie. The colonies are loosely grouped together into huge underground "cities."

In 1901, one of these "cities" in Texas was estimated to be 240 miles long and 99,400 miles across and to contain 400 million prairie dogs. Between them, they would have eaten as much grass as 1.5 million cows. Because farmers did not want to lose so much grass, they began a massive campaign of poisoning prairie dogs, so that today they are very much rarer – although there are still several million of them.

The poisoning of the prairie dogs also had a devastating effect on the black-footed ferret, which relied entirely on prairie dogs for food. A slim and agile hunter, with black face markings like a bandit's mask, it could creep into prairie dog burrows to catch them.

As prairie dogs were killed off, the number of ferrets dropped dramatically, until by the 1970s they were thought to be extinct. Then, in 1981, a small colony was discovered in Wyoming. Research by scientists showed that they were at risk from a disease more usually found in dogs. The scientists decided to take them into captivity for safekeeping. In all, 18 were caught and since then no more have been seen in the wild.

However, with careful breeding, the captive group had increased to 300 by 1991, and 49 of them were released back into the wild in an area where there were still plenty of prairie dogs. The next summer, six young ferrets were sighted in the release area – the first sign of success in a project to return the masked bandit to the prairie.

Black-footed ferret

Prairie dog burrow system

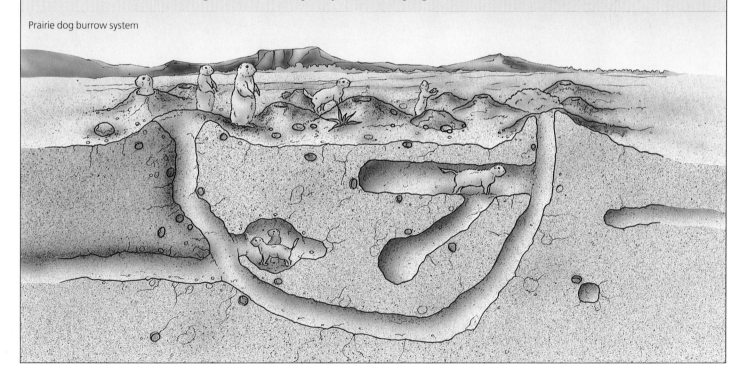

Pampas life

To the east of the Andes mountains and south of the Amazon rain forest, a huge area of South America is covered in another vast plain of grassland, called the pampas. Droughts there are frequent, and the rain comes in torrential storms, draining away quickly without penetrating deep into the soil. This stops trees from growing and allows large areas of grass to flourish.

Although South and North America are now connected by a narrow neck of land, this was not always the case. For long periods, South America was a separate continent, completely cut off from the rest of the world. As a result, many South American animals evolved in isolation and are unique to the area. It was only after the land "bridge" formed between North and South America that animals from the north could move slowly southward.

The pampas, therefore, has very different animals from those of the prairies. The only deer is the rare pampas deer, and the largest grazing animal is an ostrichlike bird called the rhea. It roams areas of tall-grass pampas in flocks of from 20 to 30, eating anything from grass to grasshoppers.

◁ The greater rhea is a flightless, ostrichlike bird, standing shoulder-high to a human. It makes a massive nest on the ground, in which the female lays up to 30 eggs that are about twice as big as a hen's eggs.

The main grazers of the pampas are small mammals. These include the cavy – the wild ancestor of the guinea pig – and the mara, which is sometimes called the Patagonian hare because of its long ears and bounding movements. Both have gnawing teeth, which are typical of all animals called rodents. This group also includes rats and mice.

Hunting these rodents is the maned wolf, which, despite its name, is actually a long-legged fox. Other hunters include the pampas fox and the caracara, a large falcon that rarely flies but hunts and scavenges on the ground instead.

But actually, the main grazers of the pampas are the insects. There can be as many as 60 surface-living insects in an area of pampas the size of this page, including grasshoppers, beetles, ants, and butterflies and their caterpillars.

Termites are among the strangest insects found in the pampas and other grasslands around the world. Antlike in appearance, termites live in huge colonies in towering fortresses made of mud and

▷ The maned wolf of the pampas does not need its long legs to catch its prey of cavies and other small mammals. Probably their main use is to help it see over the tall grass.

chewed wood. They feed on dead wood and other plant material, sometimes demolishing whole living bushes.

With so many insects in the pampas, there are plenty of insect eaters. The giant anteater, despite its name, feeds mostly on termites and can eat as many as 30,000 in a day. Armadillos also eat termites, ants, and many other insects.

Savanna spectacle

As well as the steppe, prairie, and pampas, other countries such as Australia have large areas of dry grassland. But perhaps the most famous of all grassland areas is the savanna of Africa. It stretches from the Sahara almost to the southern tip of Africa, only replaced by rain forest in West Africa and desert in parts of the south. The savanna is not all grass; in places it is mixed with thorny scrub or open woodland.

It is on the savanna that the huge herds of game animals roam: the gazelles, wildebeest, and zebras that provide one of the world's greatest wildlife spectacles. Among them are smaller numbers of giraffes, buffaloes, rhinoceroses, and elephants, all feeding on the grass or the leaves of trees and bushes. The game animals are hunted by lions, leopards,

◁ As their name suggests, armadillos, like this nine-banded armadillo, have a flexible armored shell. They can roll up into a tight ball to escape hunters. They feed on insects, including ants and termites.

cheetahs, hyenas, and hunting dogs.

Also feeding on the grass are ostriches and large numbers of rodents and hares. These are preyed on by snakes, small cats, mongooses, weasels, and their relatives called civets. Insects abound among the grass and are fed on by storks, egrets, and a relative of the mongoose called the meerkat. When it is winter in the Northern Hemisphere, these insect eaters are joined

◁ The meerkat, a type of mongoose from Africa, lives in colonies of from 10 to 15 animals. Outside their burrows, they share the duty of lookout, watching for enemies such as jackals and eagles.

▷ Few animals are strong enough to break into termite nests to feast on the insects inside. Both the giant anteater of South America (top right) and the aardvark of Africa (bottom right) have strong claws to open nests, and long snouts and tongues to suck up the termites. The name aardvark means "earth-pig" in Afrikaans, a language of South Africa.

by swallows, martins, and warblers that migrate from their breeding grounds in Europe.

Dotted across the African savanna, as on the pampas, there are massive termite mounds. When the termites venture out to feed, they are preyed on by lizards, snakes, and other insect eaters, but only two strange creatures, called the aardvark and aardwolf, are strong enough to break into the mounds to feast on termites.

HONEY PARTNERS

Honeyguides are birds related to woodpeckers and live in the bushier parts of the African savanna. They feed on insects, which they catch in flight, but they have a particular liking for the larvae and wax from beehives. The birds' tough skin protects them from bee stings, but they are not strong enough to open a beehive without help.

When a honeyguide finds a suitable beehive, therefore, it flies off in search of a ratel (honey badger). The bird attracts the ratel's attention by chattering loudly and fanning its tail, then flies toward the hive. The ratel follows and breaks into the hive with its strong claws. It feasts on the honey inside but leaves the larvae and wax for the honeyguide.

Humans have also learned to take advantage of this partnership, following the honeyguide to a beehive to collect honey, but always leaving plenty of food to reward the honeyguide's efforts.

◁ △ Termites build massive mounds as tall as lamp-posts, complete with air vents. Up to 2 million worker termites live inside the mound, but there is just a single queen (above), who swells up to the length of a finger as she lays over 30,000 eggs a day.

Unnatural grasslands

Strangely, while human activities have destroyed natural grasslands in many parts of the world, they have also created huge areas of grass in other places. When trees are felled and sheep or cattle are allowed to graze in large numbers, an artificial grassland is created. This is now the main land type over vast stretches of Europe and the United States.

Because it is newly opened and heavily grazed, this grassland usually supports far fewer flowers and insects than natural grassland, and therefore fewer wild grazing animals and insect eaters. Hunting animals are often regarded by farmers as a threat to their livestock, so they are frequently shot or poisoned.

Nevertheless, a few animals of the steppe and prairie can find a temporary home on these artificial grasslands. However, if grazing animals were removed, tree seeds would soon start growing and the natural woodland would begin to return.

In places with thin soil over chalk or limestone rock, deep-rooted trees have never been able to grow, and a more natural type of grassland can be found. This is often full of attractive flowers such as daisies, orchids, cowslips, and rock-roses. Many butterflies feed on these flowers, and the grass is often alive with grasshoppers and crickets. Even there, though, grazing by sheep or cattle is essential to stop dense mats of rough grass or scrubby bushes from spreading and from smothering the more attractive open grassland plants.

△ The pasqueflower is an attractive flower found on lightly grazed chalk and limestone grassland across Europe.

▽ The stone curlew is found on open grasslands and semi-desert across southern and central Europe and Asia. A few pairs also manage to breed on plowed fields – provided the crops are not harvested before the young have left the nest.

THE LIVING DESERT

Few places on Earth cause more problems for living things than deserts. They are hot, dry, and either bare and rocky or covered in shifting sand. Yet, against all the odds, a few plants and animals survive there.

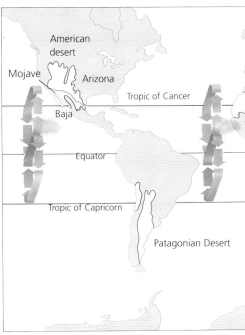

American desert

Mojave

Arizona

Tropic of Cancer

Baja

Equator

Tropic of Capricorn

Patagonian Desert

Although places like the Antarctic are, strictly speaking, cold deserts, the term desert is normally used only for places that are hot and dry. Deserts are usually defined as places that get less than 10 inches of rain a year. Often, though, this rainfall comes in a few brief storms, and some deserts only receive rain once or twice in a century.

Hot deserts cover one-fifth of the Earth's land surface. By no means are all of them covered in shifting sand dunes; most are bare and rocky.

The making of a desert

Today's deserts may not always have been desert. Evidence suggests that the Sahara, for example, was dry savanna just a few thousand years ago. A gradual shift in climate caused it to change to desert, although in recent years overgrazing by goats and sheep has made the edges of the desert move outward even farther. However, desert regions were always fairly dry, and they have shifted between being grassland and desert as the ice ages have come and gone nearer the Poles.

◁ Deserts are comparatively lifeless, although a few large mammals, like this oryx, wander over them. On average, the total weight of animal and plant matter in any patch of desert is only about one-fortieth of that in the same area of savanna, and less than one-thousandth of that in a rain forest.

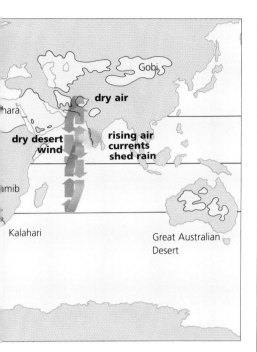

△ Dry winds spreading out from the equator create deserts around the Tropics. These wind patterns explain why there are often pairs of deserts north and south of the equator.

The reason for their dryness can be explained by their position on the globe. The Earth rotates around the sun in such a way that the sun is always overhead somewhere in the broad belt around the planet called the tropical zone. This zone bounded by the Tropic of Cancer in the north and the Tropic of Capricorn in the south. The middle of the tropical zone, marked by the equator, has the sun overhead for longest. As the land at the equator is warmed, it heats the air above it. The hot air rises, carrying with it moisture from the sea and the land. As this air rises, it cools, and sheds its moisture in the year-round rainfall that creates the tropical rain forests.

The hot air then flows toward the Tropics, 930 miles north and south of the equator, but by now it has lost all the water it once carried and so it brings no rain to the land. It then falls toward the land near the Tropics and is sucked back toward the equator, creating hot, dry winds over the desert surface for most of the year. These winds draw moisture from the land, making it drier still.

Because the desert air is dry, it does not form clouds, so the sun beats down mercilessly on the land, and temperatures, even in the shade, can rise as high as 136°F.

DESERT SURVIVORS

Despite their large size, camels can survive for long periods in the desert, wandering in search of food and water. Thick wool on their backs shades them from the sun, while the rest of their bodies are almost naked in order to lose heat quickly. Their body temperatures can vary widely without harm and, to conserve water, they do not sweat until their temperature reaches 104°F. Their urine is also highly concentrated to reduce water loss.

Although they inevitably lose some water in the heat, camels can survive even when they have lost water equivalent to a quarter of their body weight. Humans would not survive half that degree of water loss, because a shortage of water makes our blood thick and sticky. This prevents it from circulating properly, causing rapid overheating and death. Camels, however, can remove water from their body fat to keep their blood thin enough for proper circulation.

Camels do not store fat beneath their skin, as most mammals do, because this would block heat loss. Instead, they store up to 99 pounds of fat in their humps. This can be broken down when needed, to provide energy and release about 13 gallons of water.

▷ The single-humped Arabian camel, or dromedary, has been domesticated for 6,000 years. Although their wild ancestors have been extinct for 2,000 years, some have returned to the wild in various parts of the world, including Australia.

△ Only about 500 wild, two-humped Bactrian camels still survive in the Gobi Desert in Asia. They are well adapted to life there, having nostrils that can close in sandstorms and webbed feet for walking on soft sand.

After a long period without drinking, camels become thin and their humps shrivel. However, when they find water, they can drink 24 gallons in 10 minutes and are soon restored to their normal appearance.

However, without clouds, temperatures fall rapidly at night, sometimes to well below freezing.

Life in the hothouse

Few plants and animals can cope with the daytime heat and the huge drop in temperature at night, so deserts are comparatively empty of life. However, some highly specialized plants and animals do live there.

Most animals survive the heat by avoiding it. They spend most of the day under stones or in burrows and come out to feed only at night or in the cool of dawn or dusk. The animals that do venture out in the heat of the day rely on a whole range of adaptations to keep cool. Some, like the camel and ostrich, can allow their body temperature to rise and fall by a certain amount without ill effect. Many have light-colored coats to reflect heat. Some have large ears, which help them to lose heat, while others pant to cool down. The ground squirrel of the Kalahari Desert uses its bushy tail like a parasol to shade itself.

But above a certain temperature, even the most highly adapted animals cannot cope. In the hottest deserts, nothing moves in the heat of the day.

▽ Desert scorpions, relatives of spiders and ticks, get all the moisture they need from the insects and spiders they hunt with the aid of their stinging tails.

Precious water

The problem for desert animals is that most ways of keeping cool also involve losing water. Panting, sweating, even dribbling urine over their legs, as tortoises do, all use a lot of water. Yet for most animals, the only major source of water is the food they eat.

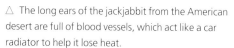

△ The long ears of the jackjabbit from the American desert are full of blood vessels, which act like a car radiator to help it lose heat.

The secret of survival in the desert is the careful saving of water. The hard, horny covering of desert scorpions, spiders, and many insects, and the scales of reptiles, are all waterproof and help to prevent water loss.

The breath of animals that spend the day underground creates a moist atmosphere in their burrows, helping to reduce water evaporation from their bodies. Gerbils in the African deserts do not waste even this moisture. They store dry seeds in their burrows when they sleep. These absorb moisture from the air, and when they wake, the gerbils eat the seeds, regaining some of the water they had lost earlier through their breath.

The rapid drop in desert temperatures at night causes dew to form, because cold air cannot hold as much moisture as warm air. This is an important source of water. Small insects such as ants drink the dewdrops directly, while larger animals feeding at dawn take in dew with the plants they eat.

Some deserts, such as the Namib in Africa and Baja in Mexico, are near the

Sand barriers

In a few deserts, the constant heating and cooling has shattered the rock into tiny sand grains, which then blow in the wind to form massive sand dunes. This sand can often become uncomfortably hot, so that lizards have to shuffle from foot to foot to keep it from burning them and insects have to lift themselves on long legs above the scorching sand.

The tiny, shifting sand particles also make movement difficult, but again some animals have ways of overcoming this. Many desert lizards, for example, have bristles, hard pads, or webs on their feet to help them walk over the surface and dig into the sand. The sidewinder snake wriggles over the sand in a series of sideways S-bends, yet it can easily move faster than a human across the sand.

Just a few inches below the surface, sand remains cool. Many animals take advantage of this by living underground. Tunneling is difficult, however, because the smooth, dry sand soon caves in. Instead, many lizards and snakes "swim" through the sand. Some lizards have lost their legs to

△ The roadrunner of Arizona and Mexico runs swiftly to catch lizards, snakes, and scorpions. It is able to extract water from its digested prey, and when it feeds its chicks, it dribbles this water from the back of its throat for the chicks to swallow.

▽ Grooves along the body of this fog-basking beetle help it collect water from the fog that sweeps over the Namib Desert.

coast. As the land cools at night, fog rolls in from the surrounding warm sea. Some plants are able to absorb this moisture, but it is of little value, because it never penetrates the soil and rapidly evaporates with the morning warmth.

The fog is not wasted on the fog-basking beetle of the Namib, however. The beetle leaves its sand burrow at dawn and climbs to the top of a sand dune. It lowers its head and raises its back in the air so that droplets of water form on its body from the fog and trickle down grooves along its side into its mouth. People in the Baja Desert are now trying to copy the beetle, using special nets to gather moisture from the desert fog.

However, most large desert mammals and birds, such as the camel, oryx, kangaroo, and ostrich, must spend their entire lives wandering in search of water. A bird called the sand grouse, for example, can fly 50 miles to find water. The male has special feathers on its chest that rapidly absorb and hold water. It then returns to the nest, where the chicks suck the feathers to get a drink.

allow easier movement through the sand. Their eyes are protected from the sand by a transparent cover.

Some desert spiders solve the problem of collapsing sand by lining their burrows with a silken web. Some also build trap-

doors at the mouth of their burrows, which they use to catch passing insects.

An even more unlikely underground creature is the naked mole rat (which is more closely related to a rat than a mole). It uses its large teeth to tunnel through the

△ The sidewinder snake can move at amazing speeds across the desert sand in a series of sideways S-bends. Only two parts of its body touch the ground at any one time, protecting it from the hot sand.

sand and to gnaw at the roots and bulbs of plants for food. It lives in underground colonies of up to 100 animals, but only one female in the colony can reproduce. The others serve as workers or as "soldiers," which can even fight off snakes with their long, sharp teeth.

Drought plants

Many desert plants, including cacti, survive by storing water in their fleshy stems. However, there are other ways to survive the drought. Some desert plants have roots that reach deep underground to hidden water supplies. Others have roots that creep widely just below the surface to collect all the available dew and absorb water as quickly as possible when it rains. The creosote bush of the American desert forms such a thick mat of roots that no other plant can grow within several feet of it.

◁ The naked mole rat lives beneath the sand of North African deserts, coming to the surface only at night. It has virtually no eyes, and is almost hairless.

LIVING WATER TANKS

survive long periods of drought. One giant cactus lived for six years without water in a laboratory.

However, cacti must also take full advantage of the rain when it comes. Most have threadlike roots. These spread widely near the soil surface so that the plant can absorb the maximum amount of water in the torrential thunderstorms that often break the drought.

Although true cacti are found only in the deserts of the United States, some, such as the prickly pear, have been introduced to Africa. However, a number of totally unrelated African plants have developed (or evolved), quite separately, with similar adaptations for desert survival and they look confusingly like cacti.

Cacti are among the most well adapted of all plants to desert life. They have largely lost their leaves to reduce water loss, and they photosynthesize instead through their stems and branches. These are swollen, providing the cacti with the biggest volume of fleshy tissues in which to store water, yet the least surface area through which water is lost.

Their outer skin is thick and waxy to reduce water loss, and their stomatal pores (see page 51) are sunk in deep pits that hold moist pockets of air, further reducing evaporation. Some cacti are covered in silky hairs. This also helps to

prevent water loss. In many cacti, leaves are replaced by thorns or spines to discourage grazing animals. Some, like the saguaro of the American Southwest, are also poisonous.

Thanks to these adaptations, cacti can

▽ These "living stones" are in fact the fleshy leaves of a plant, hidden beneath the ground. It is called a pebble plant or stone cactus (although it is not a true cactus). Its appearance may protect it from grazing animals while also helping to save water.

▽ Although cactuslike in appearance, this African plant belongs to a quite different family, the euphorbias, which are related to the spurges that grow as weeds in farmland.

△ The strange welwitschia plant of the Namib Desert has a swollen root, like a massive turnip, and just two leaves grow from this. These fray and wear at the ends but can go on growing for a thousand years.

The welwitschia plant in the Namib Desert of Africa is adapted to take advantage of the sea fog there. Its huge leaves are frayed and tattered at the ends, but on the leaf's surface are many pores through which fog or dew passes. Just beneath the leaf's surface are fibers that soak up the moisture like a sponge.

The plants that are best adapted to last through the drought periods, however, are in the form of seeds, bulbs, or tubers underground, waiting for rain.

Life-giving rain

When the drought eventually breaks and the rain comes, the desert is transformed overnight. Plants that have survived only as seeds begin to sprout. Within the space of a few days, they must flower and set seed, ready for the next drought.

The rain triggers the withered seed heads of other plants to open suddenly, catapulting out their seeds. But the seeds would not grow successfully after just a

▽ Masses of spadefoot toad tadpoles hatch in the muddy pools that briefly form when the rains come to the Arizona desert.

passing shower. They need a prolonged, drenching rain. The seeds of some plants, therefore, contain a chemical in their outer coats that keeps them from sprouting until there has been enough rain to wash the chemical away. By then, the ground will be moist enough for them to grow successfully.

The arrival of rain allows the plants that have survived on stored water to top off their reserves. These plants also produce their flowers now, because pollinating insects also come to life with the rain.

The rain also triggers the appearance of animals that have barely been alive during the long dry season. These include tiny fairy shrimps, which survived the drought as eggs. Now, within a few days, they hatch, mate, and lay their eggs in the temporary pools. When the drought returns, the eggs can survive in the dried mud for up to 50 years. They are so tiny that they can sometimes blow over 60 miles in the wind, helping to spread the fairy shrimp to new areas where pools might form.

Even more remarkably, spadefoot toads emerge with the rains in the deserts of Arizona, after having spent 10 or more dry months underground using up their reserves of food and water. The male toads quickly gather in the temporary pools and call to attract females. They mate and lay eggs immediately, then leave the pools to feed on insects before the drought returns.

The eggs hatch into tadpoles, which grow quickly. Some feed on algae in the pools, while others hunt fairy shrimps. Either way, they must be ready to turn into adult toads within three weeks. By then, the pools will be drying, the brief flowering of the desert will have passed, and the long drought will be starting again.

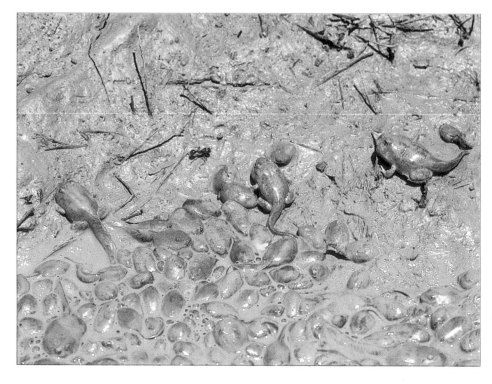

▷ When the rains come to a desert like the Mojave, buried seeds burst into bloom, producing spectacular displays of flowers for a few days before the drought returns.

In the Rain Forest

Tropical rain forests, often called "jungles," are crammed with life from the tops of the tallest trees to the dark forest floor. The great variety of plants and animals may result from ideal living conditions in the forest – or from a fierce struggle to find living space.

In 1832, the great naturalist Charles Darwin was one of the first Europeans to explore the jungle of Brazil. He wrote in his journal, "Delight is a weak term to express the feelings of a naturalist who for the first time has wandered by himself in a Brazilian forest," and added that "nothing but the reality can give any idea how wonderful, how magnificent the scene is."

Jungles are known more correctly as tropical rain forests (they are often not as dense and tangled as the name "jungle" suggests). As Darwin had seen, they are incredibly rich in wildlife. For example, the rain forests of South America are home to 30,000 species of flowering plant, and an estimated 30 million species of insect live in the world's rain forests. Altogether, although rain forests cover less than 7 percent of the world's land surface, they hold at least half the world's species of plant and animal.

Much of this life lives high in the trees, hidden from anyone walking on the dark, damp forest floor. In fact, the forest is like an apartment building, with many different plants and animals living on each level. This multi-story existence of life partly explains the richness of the forest, but the real key is the climate.

Rain and cloud

The name "tropical rain forest" perfectly sums up the weather that creates the forest. All rain forests lie close to the equator, where days are uniformly 12 hours long and the climate is warm throughout the year. Typically, the average temperature will only range between 73°F and 87°F throughout the year.

The warmth of the land heats the air above, causing it to rise and shed its moisture as rain. Most rain forests, therefore, have at least 98 inches of rainfall a year, and some have twice that amount.

This wet, warm world with plentiful sunlight is perfect for plant growth, so dense forest flourishes. The trees grow and flower throughout the year and have evergreen leaves.

◁ Rain forests have at least 98 inches of rain a year and often much more. Water from the trees adds to moisture in the air, so that clouds and mist hang over the forest. Torrential rainstorms are frequent, often accompanied by thunder and lightning.

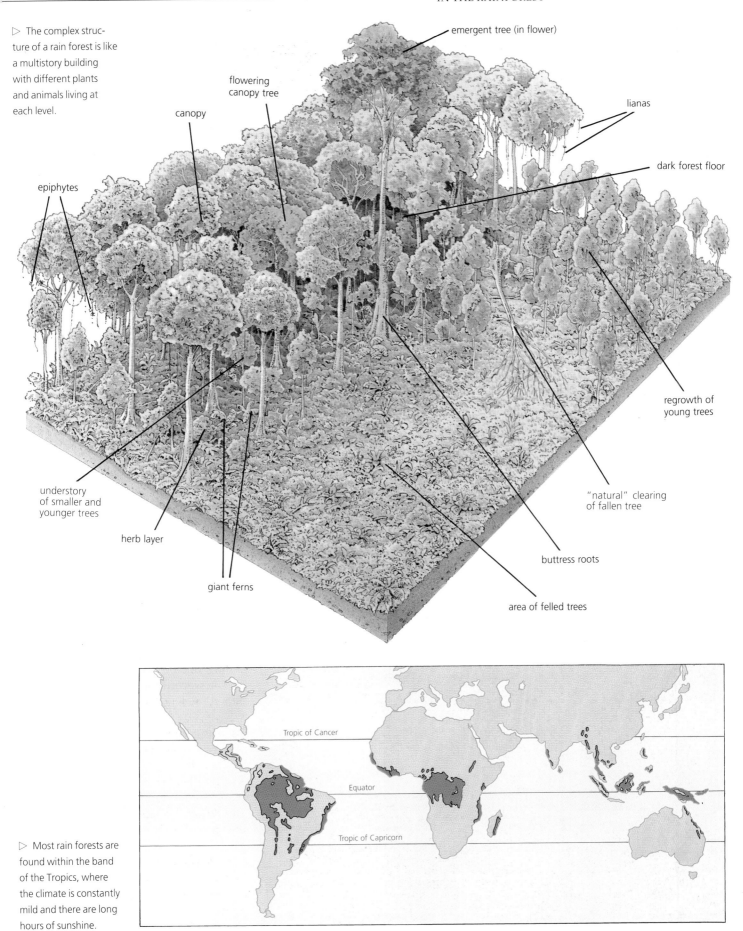

▷ The complex structure of a rain forest is like a multistory building with different plants and animals living at each level.

emergent tree (in flower)

flowering canopy tree

canopy

lianas

dark forest floor

epiphytes

regrowth of young trees

understory of smaller and younger trees

"natural" clearing of fallen tree

herb layer

giant ferns

buttress roots

area of felled trees

▷ Most rain forests are found within the band of the Tropics, where the climate is constantly mild and there are long hours of sunshine.

Tropic of Cancer

Equator

Tropic of Capricorn

▷ As much as two pounds of dead plant and animal material can fall on every square foot of forest floor from the trees above during a year. This provides food for many decomposers, such as this fungus.

The trees themselves also affect the climate. They gather water from the soil and pass it out through their leaves through the process of transpiration (see page 51). This further moistens the air, so that clouds form and hang over the tree-tops like smoke. This blanket of cloud protects the forest from the baking daytime heat and the nighttime chill of nearby desert regions, keeping temperatures within the perfect range for plant growth.

Rain forests slightly farther away from the equator remain just as warm, but they have a dry season of three months or more when little rain falls. Many trees shed their leaves during this dry season, and grow new leaves when the wet season, or monsoon, begins – hence, the name "monsoon forest" for these areas.

Another type of rain forest grows on tropical mountains. It is often called "cloud forest" because clouds constantly hang over the trees, like fog.

Jungle giants

The warmth, sunlight, and moisture allow trees of many varieties to flourish in the rain forest. Most are between 130 and 164 feet tall, and their leafy branches interlock to form a continuous, high-level expanse of green, called the canopy. A few taller trees, up to 196 feet tall, tower above the canopy. They are called emergents, because they emerge above the surrounding sea of green. Often there are shorter trees in the gaps beneath the canopy, creating leafy layers below.

All these trees rely for their survival on tiny organisms in the ground. The soil in rain forests is often surprisingly poor in nutrients. However, the mild, damp conditions are perfect for a variety of insects, bacteria, and fungi that live as decomposers. These rapidly break down all the dead plant and animal material that falls to the forest floor, releasing nutrients and minerals in a form that is soon absorbed by the dense network of tree rootlets spreading through the soil. The decomposers work so well in the rain forest that a leaf will disappear completely within six weeks, compared with a year for a leaf in a European oakwood.

◁ Many rain forest trees have spreading, aboveground roots, which act like props to hold the tree up. They are called buttress roots or prop roots.

▷ Monkeys, like this spider monkey, feed on the abundant fruits of the forest. They are wasteful eaters, and much of what they collect falls to the forest floor, where the seeds inside the fruits can germinate.

PONDS IN THE SKY

Bromeliads are a family of South American plants that includes the pineapple. Most live as epiphytes, perched high on the branches of rain-forest trees. Although they get some of the minerals they need from the plentiful rain, they have an extra source of nutrients. Their leaves overlap at the base to form a cup, which traps debris falling from above. This is broken down by decomposers and absorbed by hairs on the inner surface of the leaves.

These "traps" also collect water. Some can hold as much as a small bucket. Many animals make use of these treetop tanks. Worms, dragonflies, water beetles, mosquitoes, and other insects lay their eggs in these aerial ponds. Their young spend their lives there, either feeding on green algae in the water or hunting other inhabitants of the tanks. If they die, their bodies add to the nutrients available to the bromeliad.

Land crabs also use the tanks, but the most obvious inhabitants are frogs (above left). Some, such as poison-dart frogs, are brightly colored, a warning that they are deadly poisonous. Their poison is still used by Indians to tip their poison arrows.

usually first to begin growing there. When these die, their remains form a thin "soil" in hollows and forks in the branches, in which other plants can grow.

These epiphytes (literally plants that grow on other plants) get all their water from the falling rain, which also contains the minerals and nutrients they need for growth. Epiphytes are not parasites on the tree; their roots simply wrap around the branches to hold them in place, then dangle down in a thick curtain that helps to increase their water absorption.

Many orchids, ferns, cacti, and bromeliads grow as epiphytes on rain forest trees. Some of them are torn from the trees and sold in garden centers as "air plants."

For anyone walking on the forest floor, the most conspicuous plants are climbers called lianas, which hang from the trees like woody ropes. These lianas begin life as seedlings on the forest floor, but when a gap in the canopy lets the light in, they shoot up and attach themselves to a young tree. They then grow up with the tree, attached to it by roots, hooks, or coiled threads, or by twining themselves around trunks until they reach the sunlight above.

▽ More than 9,000 species of orchid grow as epiphytes in the Tropics. Many are collected from the rain forests and sold in flower shops. Very few of them can be grown successfully in greenhouses.

The struggle for life

The rain forest is the ideal environment for the growth of many different trees. As many as 200 species can be found in an area of rain forest the size of a football field. The result is fierce competition for living space.

This struggle is mainly confined to the first months of a tree's growth. Tree seedlings generally cannot grow in the shade of the forest floor, because there is not enough light for photosynthesis. However, when a canopy tree dies and falls to the ground, the sunlit gap in the forest floor is soon invaded by tree seedlings.

Some forest trees produce fruits and seeds in huge numbers. These fruits and seeds can lie on the forest floor for several years before germinating when a tree falls and sunlight reaches them. Other species produce a few large fruits with plenty of stored food to allow their seedlings to establish themselves quickly.

Most rain forest trees rely on animals to eat their fruits and spread their seeds. Some fruits are dropped by the animals, spreading the seeds. Others are eaten, and the seeds pass undamaged through the animal's digestive system and are passed out in its droppings.

Plant hitchhikers

Although few plants can grow on the dark forest floor, the regular rainfall allows many plants to grow as "hitchhikers," or epiphytes, on the branches of trees high above the ground. Mosses and ferns are

◁ This centipedelike creature, called peripatus, comes out at night or in rainstorms to feed on dead animal matter. It lives in the rain forests of Africa, Asia, and South America. Similar animals have been found as fossils from 500 million years ago.

Life on the floor

A constant rain of dead plant and animal matter falls to the forest floor, providing food for the many different decomposers. Termites are particularly abundant and can make up almost three-quarters of the weight of invertebrate animals on the forest floor.

The constant dampness on the rain forest floor allows animals normally associated with water to live there. Leeches,

▽ The Javan rhinoceros is one of the shiest and rarest of rain forest animals. Only about 50 are left in remote corners of Java, in Indonesia, because of hunting and forest clearance. This is a female; the male has a slightly bigger horn.

THE JUNGLE STRANGLER

The strangler fig has a remarkable way of finding space for itself in the crowded rain forest: it literally squeezes out another tree. The fig begins life as a seed dropped by an animal on a branch high in a tree. When the seed germinates, it sends out

roots, which grow down perhaps 98 feet to the forest floor.

When the roots reach the ground and begin to take up water and nutrients, the fig grows rapidly. It forms a large, leafy crown, which competes with its host tree for light. Its roots thicken and spread until they form a woody cage around the host tree.

Eventually, perhaps a century later, the host tree is smothered by the fig and dies, but by now the fig is a free-standing tree held up by a cylinder of woody roots.

Now the fig produces its flowers. These are contained within a swollen, green, hollow structure. They are pollinated by wasps, which lay their eggs in the young flowers and live most of their lives in the developing fruits. This relationship is so close that every species of fig has its own species of fig wasp.

After pollination, the fruits develop – up to 100,000 of them from a single tree. These fruits are an important food for many species of bird and mammal. Each fruit contains dozens of hard, gritty seeds. These seeds pass undamaged through the animal's digestive system and are deposited in its droppings, perhaps on a suitable branch for a new strangler to begin growing.

◁ Some rain forest frogs avoid the need for ponds by producing eggs surrounded by a thick jelly coat. They lay these among mosses on the forest floor, where they will not dry out. The tadpoles develop into tiny frogs while still inside this jelly.

▷ Flowers like this, which have evolved to be pollinated by fruit bats, produce a musty smell at night to attract the bats. The flowers usually hang clear of other vegetation because bats are not agile fliers.

flatworms, and even land crabs are common, along with large numbers of frogs and toads.

Seedlings in clearings and falling fruits and vegetable matter provide food for grazers, including a variety of deer, antelopes, pigs, and piglike creatures called tapirs. In South American rain forests, long-legged rodents called agoutis and pacas feed mainly on fruits. Elephants and buffaloes live in Asian rain forests, as well as two very rare species, the Javan and Sumatran rhinoceroses.

In the canopy

Most rain forest animals, however, live in the bustling world of the canopy. There flowers, fruits, and leaves provide abundant food throughout the year and the branches form a convenient, high-level walkway.

The fruit eaters include a wide variety of monkeys, squirrels, fruit bats, and birds. The strong beaks of parrots and macaws allow them to crack open hard nuts.

Other birds, such as hummingbirds, sunbirds, and honeycreepers, feed on nectar from flowers. They hover in front of the flower with fast-beating wings and use their long, curved beaks to reach deep inside the flower for the nectar. In the process, they may pollinate the flower.

Most bird-pollinated flowers are red in color to attract birds. They are usually tubular in shape and grow clear of other vegetation, to allow the birds easy access to the nectar. Some bats also feed on nectar, and flowers pollinated by them typically produce a rather musty smell during the night. Pollination by birds and bats is common only in tropical regions, because only there do animals have a guaranteed supply of nectar throughout the year. However, in the rain forest, the most common pollinators are beetles, bees, wasps, and butterflies.

The main food source in the rain forest is leaves, both from the trees and from the epiphytes that grow on them. They are eaten by many monkeys, squirrels, and sloths, and by large numbers of insects. Because leaves are difficult to digest, leaf eaters need large, highly specialized digestive systems.

Plants benefit from animals that take their nectar (and accidentally pollinate the flower) or eat their fruits (thus spreading their seeds), but they suffer only harm when animals eat their leaves. Many plants, therefore, produce chemicals to discourage grazers. Some animal species have responded, over thousands of years, by becoming immune to the effects of these chemicals, and so they only eat the leaves of these particular species. This chemical warfare has encouraged new species to evolve, further adding to the great variety of life in the rain forest.

▷ Several rain forest animals have webs of skin between their legs that allow them to glide between trees. These animals include "flying" lizards, squirrels, and, most skillful of all, the colugo or flying lemur (shown right), which can glide almost 200 feet between trees.

At any time in the rain forest some trees will have fruit or be in flower, but they will often be far apart. Animals that rely on flowers or fruit for food, therefore, need to move widely in search of a meal. This is not a problem for birds or bats that can fly, but for monkeys it means they must be very swift and agile to move through the branches. To help, some South American monkeys have a prehensile (grasping) tail, which acts like an extra arm or leg and speeds up their movement.

Some squirrels and lizards move between trees in another way. They have a flap of skin between their legs that allows them to glide down from high up in one tree to the trunk of a neighboring tree, saving a long climb to ground level. Some snakes can move in the same way by flattening their body and spreading their rib cage to help them glide.

All the apes – gibbons, chimpanzees, orangutans, and gorillas – live in the rain forest, eating leaves and fruits. Large adult male orangutans and gorillas can become so large and heavy that trees will not support their weight. They have to live on the ground, moving between clearings where there is plenty for them to eat.

Leaf eaters in general do not need to move far in search of food and can afford to be slow-moving, like the sloth. Similarly, many insects spend their young or larval stage grazing on a single tree. Only when they hatch into adults do they fly off to find other trees on which to lay their eggs, ensuring an abundant supply of leafy food for their young.

With so many insects, there are plenty of insect eaters in the canopy. Frogs, toads, and lizards are particularly abundant in the trees. Many of the frogs have suction pads on their feet to help them clasp leaves, and they are either green for camouflage or brightly colored to warn that they are poisonous. Most stay in the canopy and breed in miniponds in the leaves of trees.

◁ Although food is plentiful in the rain forest, minerals are often in short supply. Here scarlet and red-and-green macaws from Brazil eat soil from a riverbank for an added source of minerals.

△ Despite their vast size, mountain gorillas are peaceful vegetarians. Weighing up to 440 pounds, male gorillas are too heavy to climb trees, so they must find all their food on the forest floor and in clearings.

◁ A boa constrictor lurks in a tree, waiting for a squirrel, opossum, or small bird to come too close. It will then grab its prey in its mouth and crush it to death in the coils of its body. Although boas can reach a length of 18 feet, stories of them killing humans are greatly exaggerated and almost always untrue.

Forest hunters

The largest predators live on the forest floor, hunting pigs, antelope, and deer. Jaguars, tigers, and smaller cats such as the ocelot are typical of these forest-floor hunters. Other cats, such as the margay and clouded leopard, are skilled climbers, well able to catch monkeys, squirrels, and birds high in the trees.

In the canopy, snakes lie in wait, hidden by their camouflage, to grab any animal that strays too close. Hawks and eagles fly over the canopy, ready to snatch their prey from the branches.

All these rain forest animals lead complex, interlocking lives, but all depend ultimately on the climate, which supports the rich plant growth. However, all of their lives are threatened by the destruction of the rain forest by humans (see pages 139 and 140). Unless this greed and thoughtlessness can be controlled, most of the rain forest will be destroyed within 10 to 15 years, and what Charles Darwin called the "sublime grandeur" of the rain forest will be lost forever.

WATER LIFE

Only about 3 percent of all water on Earth is fresh – the rest is salty. Yet this fresh water supports an amazing variety of life, whether it is in flowing streams and rivers or in still ponds and lakes. Wetlands around water are also home for many plants, insects, and birds.

Water provides a perfect life-support system. Living cells typically contain over 75 percent water, and the main battle for many living things is to maintain this water balance. This is no problem for plants and animals living in water, since water is all around them.

Water supports the bodies of animals and holds up water plants toward the sunlight without the need for woody stems. Enough oxygen and carbon dioxide are dissolved in water to support life, and can be absorbed easily by gills or leaf surfaces.

Minerals and nutrients are also dissolved in water. Water plants can absorb these through all their underwater parts, although they still need roots to anchor them in fast-moving water. Microscopic animals literally swim through a "soup" of food, and provide an easy meal for other animals.

Because of the unusual chemical nature of water molecules, they have a particularly strong tendency to cling together. This creates a surface tension, which acts almost like a "skin" on the water's surface. Not only does this allow lightweight creatures to walk over the surface, but it holds up the floating leaves of plants, such as water lilies.

Because all life on Earth is water-based, life is most suited to the temperatures in which water remains liquid, and these are the prevailing temperatures over most of the Earth's surface. When water freezes, it expands. As a result, it floats on the water's surface, acting like a blanket over the unfrozen water below. As it expands, ice often leaves an air gap above the water's surface. This is essential for air-breathing animals beneath the ice.

Ponds and lakes

The relatively shallow, sunny water of lakes and ponds encourages a rich growth of microscopic green algae, supported by nutrients from the surrounding land. The algae provide abundant food for microscopic animals, which, in turn, are food for larger creatures, such as water beetles and insect larvae. These are eaten by fish, which are then hunted by predatory birds such as herons, kingfishers, and fishing eagles, or, in some parts of the world, by crocodiles and alligators.

The larvae of many insects, such as mayflies and stone flies, live in ponds, feeding mostly on algae. Dragonfly larvae living in ponds are ferocious hunters, so other larvae need to hide in the gravel or mud. Caddis-fly larvae build a tubular case of twigs or stones for protection.

◁ Dragonflies are fierce hunters. They patrol over water in search of smaller insects, which they catch in flight.

Pond animals can be caught using a strong aquarium net (available from pet shops). To collect the widest range of species, sweep the net slowly backward and forward through the water and brush it against water plants. River animals are best caught by holding the net on the riverbed, facing upstream. Carefully turn over stones upriver from the net, so that animals underneath are washed into the net. Always take great care when you are in or near water, especially in fast-flowing rivers.

Empty your catch into a flat, white dish filled with water. A magnifying glass will be useful in observing your catch more closely. The drawing shows animals you might find and where they live in a pond. Look carefully at your catch to see how the animals move. Can you tell how they breathe?

If a microscope is available, you can study the tiny plants and animals of the plankton using a net made from a foot of old nylon tights, attached to a stick by a wire hoop at the top, and with a jar held in the foot of the tights. Drag this slowly through the water to collect plankton, then place a sample on a microscope slide or petri dish to see what you have caught.

▷ Many different creatures live in or near ponds. Some walk on the water's surface film or hang suspended beneath it. Others shelter among vegetation or in mud on the pond bottom, hiding from hunters prowling the open water.

STUDYING WATER LIFE

▷ Isolated lakes often support their own special fish populations. For example, shallow waters around the shore of Lake Victoria in Africa are the home of small fish called cichlids (below), of which over 200 species are found nowhere else in the world. However, the Nile perch (below left), a large predatory fish, has been introduced into the lake by fishermen and now threatens the survival of the cichlids.

▷ The dipper uses its wings to "fly" under-water, against the current of fast-flowing rivers, in search of insect larvae on the riverbed.

Animals living in small, isolated bodies of water risk inbreeding (breeding only within the same small group of individuals). This is dangerous because it can cause a higher incidence of genetic defects. In addition, sudden disasters can wipe out their entire population. Insects overcome this when, as adults, they mate and then fly off to find new ponds and lakes in which to lay their eggs.

Similarly, frogs can move overland, so the adults spread out from the pond where they were born. The tadpoles of frogs and toads live in ponds, breathing by gills, feeding on algae, and growing rapidly. As they grow, they change from feeding on plants to eating other animals, and their legs and lungs begin to develop. Eventually, they are ready to spend the rest of their lives on land, although they never move too far from water.

The life in ponds and lakes depends on the balance of chemicals in the water. If the water is enriched too much by chemicals draining off farmland or by sewage, the growth of algae can get out of control, creating an unpleasant green scum on the water's surface. This prevents light from reaching underwater plants, and as the scum dies and rots, it uses up all the oxygen in the water, killing the fish.

But shallow ponds are themselves short-lived in nature. Water plants such as reeds, which grow around the edges of a pond, trap mud, raising the bed of the pond until willow trees begin to grow. Only regular trampling by animals or clearance by humans can stop this natural process by which ponds slowly turn into woodland.

River life

Few green algae survive in the fast-flowing waters of a river's upper reaches, but enough dead matter is carried in from the surrounding land to support a variety of scavengers and hunters.

In the slower-moving lower reaches of the river, plants grow anchored to the river-bed by their roots. This zone is much richer in life. One scientist has calculated that 64 million microscopic animals, weighing one-fifth of a ton, are carried in one day under a bridge over the Missouri River.

Insect larvae, similar to those in ponds, shelter among the gravel on riverbeds, and are hunted by fish and diving birds such as dippers.

Some fish, such as salmon and sea trout, breed in rivers but spend most of their lives at sea. Others, such as eels, travel to the sea to breed, but return to the rivers to live and grow. These breeding systems ensure that adult fish are not competing for food with the mass of young fish.

The fish are hunted by otters, crocodiles, and birds. Insects hatching out from the river are gathered from the air by swallows during the day and by bats at night.

Like ponds, rivers are easily damaged by pollution. Waste gases from factories and power plants, carried long distances in the wind, can end up dissolved in rivers, turning them so acid that insect larvae cannot survive. As a result, birds like dippers also disappear from these polluted rivers.

Wetlands

Around rivers and ponds, in marshes and other wet areas, collectively called wetlands, large areas of floating vegetation, reeds, and boggy woodlands are home to many other animals. Many birds breed among the reeds, for example, while a variety of insects bore their way into the reed stems.

Peat bogs (below) are formed from the partly decayed remains of peat mosses like these. The acid peat is such a good preservative that it is possible to identify the last meals of 2,000-year-old human bodies unearthed from the peat.

One type of wetland depends entirely on a moss called peat moss or sphagnum, which traps water in its leaves like a sponge. As it grows upward, the remains at its base are so waterlogged that they partially decay, releasing acid as they do so. These partly decomposed remains are called peat. Peatlands cover some 3 percent of the land's surface and are an important habitat for a wide variety of plants and insects.

The acidity of peat preserves anything that falls into it. Fragments of plants and pollen grains from deep peat bogs provide information about what grew in these areas 10,000 years ago or more, and they offer scientists a unique record of past changes in the climate and vegetation of an area.

▽ Unless controlled, reeds will naturally colonize the edge of muddy ponds, beginning a process that will eventually turn the pond into woodland. Reed beds, however, are a valuable habitat for birds like the bittern (right), a shy and superbly camouflaged relative of the heron.

KINGDOM OF THE DEEP

Almost three-quarters of our planet is covered by seawater, forming the single largest home for life on Earth. Because living in seawater is easier in some ways than living in air, a tremendous variety of animals and plants live in different ocean locations, from the sunlit surface to the deepest, darkest ocean trenches.

Seventy-one percent of the earth's surface is covered by ocean – more than twice the area of land. The average depth of water over this huge expanse is over 2 miles, so there is an enormous volume of water in the world's oceans, with creatures living at all levels in the seawater. There is even life in the deep ocean trenches, which reach down to a depth of almost 7 miles, deeper than Mount Everest is high.

Living permanently in water is very different from living in air, and has several important advantages. Seawater gives more support, so that sea-living (marine) plants and animals do not need strong skeletons to hold their bodies in shape. Both large, bulky animals and small, fragile life-forms can float freely in seawater.

△▷ The sea supports creatures of enormous size, like the blue whale, which can be 98 feet long and weigh 165 tons. Seawater also supports very delicate creatures, like this Venus girdle, which is 6.5 feet long but only a tiny fraction of an inch thick.

OCEAN ZONES

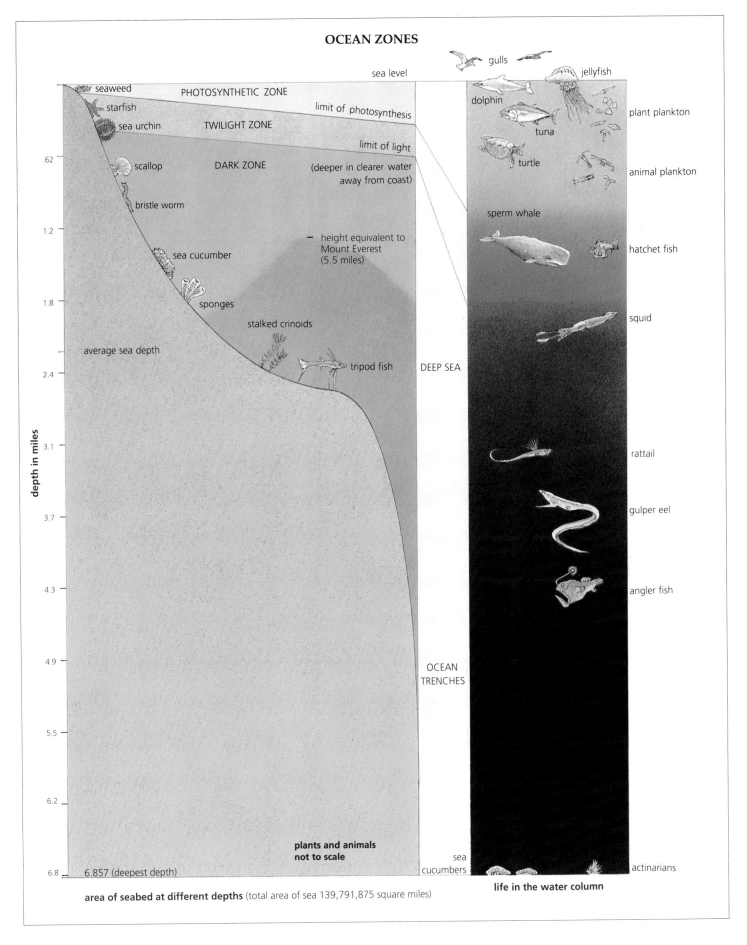

gulls

sea level

jellyfish

dolphin

PHOTOSYNTHETIC ZONE

limit of photosynthesis

plant plankton

seaweed

starfish

sea urchin

TWILIGHT ZONE

limit of light

tuna

.62

scallop

DARK ZONE

(deeper in clearer water away from coast)

turtle

animal plankton

bristle worm

1.2

sea cucumber

height equivalent to Mount Everest (5.5 miles)

sperm whale

hatchet fish

1.8

sponges

stalked crinoids

squid

average sea depth

2.4

tripod fish

DEEP SEA

depth in miles

3.1

rattail

3.7

gulper eel

4.3

angler fish

4.9

OCEAN TRENCHES

5.5

6.2

plants and animals not to scale

6.8

6.857 (deepest depth)

sea cucumbers

actinarians

area of seabed at different depths (total area of sea 139,791,875 square miles)

life in the water column

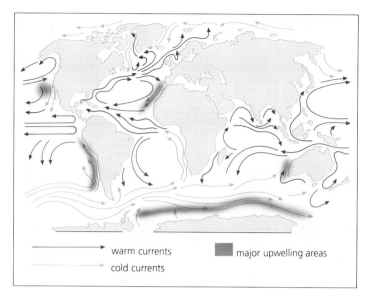

◁ The surface waters of the oceans are moved mainly by winds. Where currents meet, water is forced downward, and where currents part, water is drawn up from below. This upwelling water is richer in nutrients and therefore more beneficial for marine life.

warm currents
cold currents
major upwelling areas

△ Jellyfish tentacles can spread widely in the water, making a sticky and poisonous net for catching smaller animals. Sometimes small fish, like these whiting, live unharmed among the tentacles. They gain protection from the jellyfish, while acting as bait to attract larger fish, which are then caught by the jellyfish.

Swimming through water requires much less energy than lifting each foot against the force of gravity. Therefore, marine animals do not need to store as much energy in the form of starch and related chemicals as land animals do, and their bodies are mainly made up of proteins.

The nature of the sea

Seawater tastes salty because it has various substances dissolved in it, including a large proportion of common salt (sodium chloride). Every gallon of seawater contains around 1 ounce of dissolved solids, including nitrates, phosphates, and trace elements needed for plant growth. Seawater also contains dissolved gases, such as carbon dioxide, which plants need for growth, and oxygen, which plants and animals need for respiration. So seawater is an ideal life-support system.

Like air in the atmosphere, seawater is constantly moving, driven by winds and tides. This is very important for marine life. Water flowing past plants and animals provides them with nutrients and gases and takes away their waste products. Animals that live attached to the seabed can feed on smaller plants or animals carried past in the water. Also, the young animals and plants can be carried long distances in ocean currents to settle in new areas, just as the seeds of some land plants are spread by the wind.

Even in the deepest parts of the ocean, up to 6.8 miles below the surface, there is gentle water movement. Seawater cooled in the Earth's polar regions becomes denser and sinks to the depths, creating a vital circulation without which water in the deep sea would become stagnant, oxygen would be used up, and animal life could not survive.

Because water heats up and cools down more slowly than air, there is much less variation in temperature in the sea than on land. Therefore, ocean animals and plants do not have to cope with extremes of heat or cold or with sudden temperature changes.

Sunlight does not reach far down into the sea, because it is reflected, absorbed, and scattered by water molecules and tiny particles in the water. Some light reaches down to about 1,968 feet in clear, tropical waters, but only to about 328 feet in colder, murkier waters, especially near coasts. However, the light soon dims with depth. Only about 1 percent of the light from the surface reaches 328 feet, and this is the maximum depth at which plants can photosynthesize (see page 111). The quality of the light also changes with depth. The red part of the light is blocked out by the water within 98 feet, orange and yellow light a little deeper, and green and violet by about 328 feet. Below is a dim, blue twilight world in which the fish move as dark shadows, until total darkness takes over at even greater depths.

Near the surface, however, plants grow rapidly, providing food for animals both in the surface waters and far below.

Floating life

The free-floating and weakly swimming plants and animals that drift with the ocean currents are called plankton. The plants of the plankton (phytoplankton) are especially important because they carry out by far the greatest amount of photosynthesis in the oceans, providing energy that is then passed to animals through marine food chains.

Feeding on these minute plants are a host of tiny animals, called zooplankton. One abundant group of animals in the plankton is the copepods. Smaller than grains of rice, they eat phytoplankton and in turn are eaten by larger animals, forming a vital link in marine food chains.

Floating animals and plants have various ways of keeping themselves from sinking below the sunlit levels where the phytoplankton can photosynthesize and the zooplankton can find abundant food. Flattened shapes, spines, and hairs act like parachutes, slowing their sinking. Some can even adjust the level at which they float by altering the amount of oil, fat, or gas in their bodies. Copepods, for instance, store their food reserves as oil droplets, which help to keep them afloat.

Seaweed and sea grass

Although 95 percent of the photosynthesis in the sea is the work of tiny phytoplankton, there are important primary producers in shallow waters. In colder coastal areas, large brown seaweeds called kelps form dense forests on underwater rocks, similar in many ways to land forests.

Kelp forests are very productive, fixing several times more energy than phytoplankton in the same area of sea. But because kelps can grow only where there is a rocky seabed in shallow water, they produce only a small portion of the total energy produced by plants in the sea.

Other seaweeds are also important primary producers, especially in tropical waters. Over 90 percent of marine plants are algae (seaweeds and phytoplankton), but in some places sea grass grows on shallow, sandy, or muddy seabeds. Sea grass can produce up to 4,000 grasslike leaves in a square foot of seabed, producing as much energy through photosynthesis as a square foot of tropical rain forest.

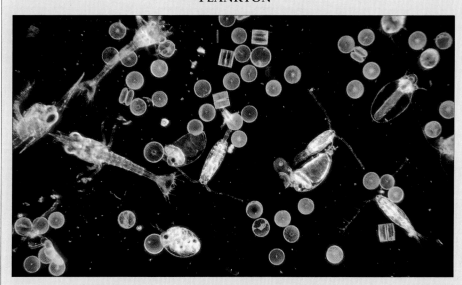

PLANKTON

Seen under the microscope, plankton have many strange and beautiful forms. The green circles in the photograph are diatoms, tiny plants with cases of a material similar to glass. (The greenish squares are diatoms seen from the side.) The tiny skeletons of diatoms sink, forming thick muds, or oozes, over vast areas of the seabed.

There are also tiny animals in the plankton. The creatures with long antennae near the center of the photo are copepods, an important food for many fish. They use their antennae for swimming and for trapping phytoplankton. Other animals in plankton include the larvae of crabs (left of center), which have bulging eyes, and cladocerans, or water fleas (such as the creature resembling a light bulb at the top right).

△ ▷ A few larger animals, such as these dugongs, eat sea grasses directly. But most of the leaves die and are broken down by bacteria and eaten by tiny invertebrate grazers. Many of the leaves are washed into neighboring areas of seabed, onto the seashore, or may even be blown inland, enriching other coastal habitats.

△ In this aerial view of a coral reef, the white line is made by waves breaking against corals on the seaward edge of the reef. The light blue behind is shallow water over a lagoon of white coral sand.

Coral reefs

In the Tropics, where sea temperatures are always about 64°F, the seabed along millions of miles of coast is fringed by shallow reefs made by animals called corals. The reefs are also home to a colorful spectacle of hundreds of kinds of fish and invertebrate animals.

There are few plankton because of the shortage of nutrients in tropical waters, so the abundance of life may seem puzzling. However, the fact that coral reefs, like seaweeds, thrive only in shallow water down to 65 or 98 feet gives a clue to their success. Although corals can feed on smaller animals using their sticky tentacles, they have an extra source of food. Their tissues contain algae, tiny plants similar to phytoplankton. These photosynthesize in sunlight. Some of the products of photosynthesis "leak" from the algae to the coral, helping it to grow.

HOW CORAL REEFS GROW

Corals are animals related to sea anemones. Corals form shallow reefs in tropical coastal waters. At first, the corals grow singly on the seabed, but in time new corals start to grow on the hard skeletons of dead ones, forming underwater coral mounds. These mounds join up sideways, making patches of reef.

At the seaward edge of the reef, strong water movement brings oxygen and plankton for food, and removes silt. There, corals grow vigorously. But corals in still water nearer the coast gradually die. Their skeletons break down into coral sand, making a sheltered lagoon between the reef and the coast.

The coral reef continues to grow out toward the sea, until its base reaches a depth at which there is not enough light for corals to grow. Other animals attach to the steep reef wall, making the most colorful spectacle in the oceans. Hard, pink, encrusting seaweeds play an important part in helping to cement the reef together.

◁▷ Live coral (left) is made up of a thin layer of living tissue over a hard skeleton, which shows up only when the coral is dead (right). The greenish color comes from algae living in the dead coral.

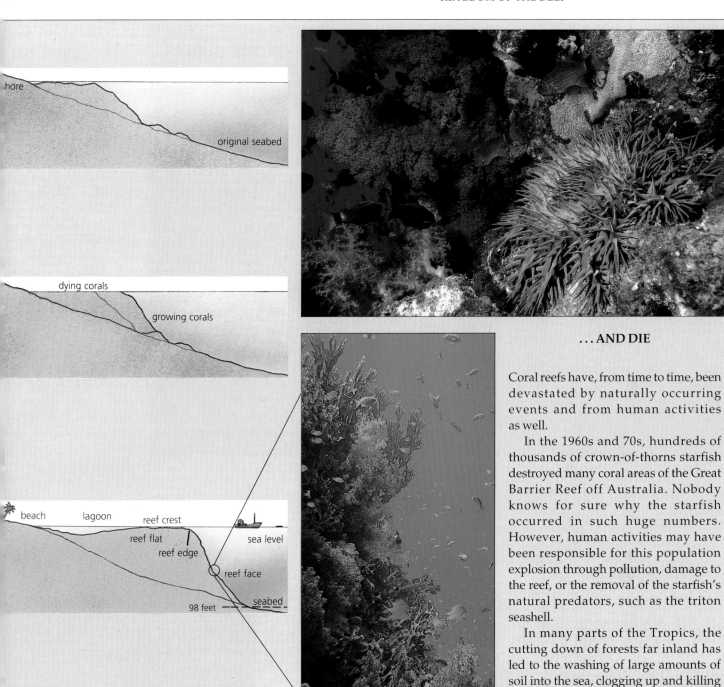

shore

original seabed

dying corals

growing corals

beach   lagoon   reef crest

reef flat   sea level

reef edge

reef face

98 feet   seabed

... AND DIE

Coral reefs have, from time to time, been devastated by naturally occurring events and from human activities as well.

In the 1960s and 70s, hundreds of thousands of crown-of-thorns starfish destroyed many coral areas of the Great Barrier Reef off Australia. Nobody knows for sure why the starfish occurred in such huge numbers. However, human activities may have been responsible for this population explosion through pollution, damage to the reef, or the removal of the starfish's natural predators, such as the triton seashell.

In many parts of the Tropics, the cutting down of forests far inland has led to the washing of large amounts of soil into the sea, clogging up and killing coral reefs.

Coral reefs grow only in shallow water because the algae need light. In turn, the algae can use coral waste products – for example, carbon dioxide and nitrogen compounds – for their own growth. Therefore, most of the primary production of the coral reef happens inside the coral, rather than in phytoplankton floating in the seawater. This means that little is lost from the system, and allows a coral reef to support abundant life.

Living on rock

Unlike land animals, which need to move about to find food and to reproduce, many marine animals live firmly attached to rock throughout their lives. They can do this because food is carried to them in ocean currents, and they need only stretch out their tentacles or feathery arms to catch it. Others find food by filtering it from seawater, which they pump through their bodies. Marine animals can also use the

water movement to carry their reproductive cells for breeding or to spread their young. So successful is this fixed way of life that there is rarely any available space on underwater rocks.

On rocks in shallow water, seaweeds are often the most obvious living things, with kelp forests in colder waters and smaller seaweeds in the Tropics. The seaweeds provide food for grazing animals, especially sea urchins and limpets, and shelter for a host of small

◁ This rock is covered with animals using the water movement to feed. Feathery structures, like the arms of the delicate pink and white featherstars, which are scattered all over the rock (to the right in the photograph) and the fluffy, white soft corals (center) trap plankton from the water. The object that looks like an orange peel (top left) is a soft coral that is closed up. The gray seasquirts (like the two seen at bottom left) pump water through their bodies, filtering out the food. The well-named sunstar (near the center) feeds in quite a different way: it eats other animals attached to the rock.

worms, brittle stars, and shrimplike crustaceans. Many of these hide away by day to avoid being eaten by fish, and come out only in darkness to search for food.

Below about 550 feet in even the clearest waters, the light becomes too dim for plants to photosynthesize. There, only animals grow, although some of them look remarkably like plants. Their names can also be confusing: sea anemones and sea firs, for instance, are animals, not plants! These animals are preyed on by

beautiful sea slugs. Although these are slow-moving, this is not a problem when their food is fixed to the rock and cannot escape.

Life in sand and mud

At first glance, a diver underwater might see little marine life on a muddy and sandy seabed, compared with the profusion of animals and plants on rocks. But holes and mounds on the seabed give clues to the

many animals hidden safely in burrows and tubes in the mud.

Animals and seaweeds cannot fasten themselves to soft mud as they can to rocks. Because nothing grows fixed on the mud surface, the animals on the seabed

◁ This seabed may appear empty of life, but the mounds, scrapes, and holes show that there are many animals living hidden in the mud.

▷ Sand eels swim in large shoals, which may confuse predators. They can also disappear head-first into the sand to escape hunters. Nevertheless, they are an important food for larger fish and seabirds.

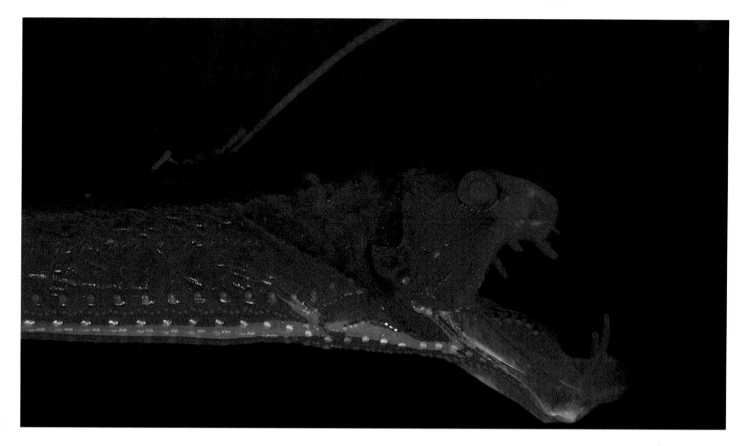

△ The Sloane's viperfish lives in the twilight levels of the deep sea. To help avoid predators, it is flattened sideways so that it is less visible from below. It produces its own light by bioluminescence, partly to show off to viperfish of the opposite sex and partly to attract prey. To catch its prey, it also uses the long, shining, threadlike lure above its head.

have no cover in which they can hide from larger predators, such as fish and crabs. At the slightest disturbance, they withdraw into their burrows or tubes. Many extend only part of their bodies out of their burrows to feed at the mud surface or from the seawater above so that they can retreat quickly into their burrows. Some emerge only at night, when there is less danger of being found by predators.

Soft mud is particularly difficult to live in, because the fine particles clog feeding and breathing parts, and there is often no oxygen below the mud surface. Relatively few, specialized kinds of animals live there, but sometimes they can be found in great numbers.

More kinds of animals live in mud that is mixed with sand and shell gravel, because this provides a greater choice of building materials for burrows and tubes. In these sandier seabeds, there are often great numbers of bivalves, such as cockles, as well as many different kinds of worms and bottom-living fish. Animals and seaweeds that usually live on rocky seabeds grow on pebbles and empty shells. Mobile animals, such as starfish and hermit crabs, can also move more easily over these firmer, sandier seabeds.

In areas where waves or currents wash away the finer mud and sand particles, the coarse sand and gravel left behind is inhabited only by heavy-shelled bivalves and a few other tough animals. The constant water movement makes these areas difficult to live in, because burrows are frequently destroyed and the moving sand wears away shells.

The deep sea

By far the greatest part of oceans lies beyond the shallow sea shelf around the continents and is too deep and dark for photosynthesis. The deep sea accounts for 85 percent of the area and 90 percent of the volume of the oceans, and is the least known and explored part of the Earth.

From 1,968 feet (or less in murky water) to the deepest parts of the oceans at 36,000 feet, there is total darkness. The seawater is

◁ A variety of animals live on the deep-sea floor, but numbers are few because little food reaches this far down. This creature is an isopod, a relative of wood lice. It is living 10,728 feet down in the ocean, a depth into which Mount Etna could be sunk almost completely.

cold (between 33.8°F and 41°F) and moves slowly because it is far below the surface winds and tides. The pressure at these depths is tremendous, but plant and animal tissues are compressed very little by pressure unless they contain air spaces.

Although the deep sea may seem a cold, dark, and inhospitable place, living conditions there are actually very steady and hardly change with the seasons. Many different kinds of animals have been able to adapt to these conditions.

A little light does penetrate the upper part of the deep sea, so that it is not totally dark. This twilight zone is inhabited by animals with well-developed eyes. Many also have organs that can produce light by chemical reactions inside their cells and tissues – a process called biolumines-cence. In this zone, the fish tend to be black and the crustaceans red (red cannot be seen in the bioluminescence). Many of the animals in this zone migrate upward at night to feed on the plankton above,

and downward into the dark during the day to avoid being seen by predators.

Below this, there is total darkness and fewer animals. Those that do dwell there are often white and have shrunken eyes that can detect little more than light and dark. Some may still have bioluminescent organs. Many fish in this zone have huge mouths with backward-pointing teeth so that they can catch large prey in these dark depths.

Strong swimmers

Perhaps the most familiar of all sea creatures are fish, whales, and turtles, which over the centuries have been important sources of food for humans. They are strong swim-mers, able to travel long distances between feeding and breeding grounds. They are not dependent, as are plankton, on ocean currents, although fish may still rely on currents to spread their eggs and larvae.

There are many species of fish in the

oceans. Most are carnivores, eating smaller fish or other marine animals.

Whales and dolphins, despite their streamlined appearance, are not fish but mammals that are highly adapted for living in the sea. Like all mammals, they need to breathe air, but they can spend up to 40 minutes underwater between breaths. They have a thick layer of fat (blubber) to keep them warm, their limbs and tail have become modified for swimming, and they are able to dive to great depths.

Whales and dolphins are truly marine; they travel the open oceans and never need to return to land, even to have their young, which are born and reared at sea. Sea cows (dugongs and manatees) usually have their young in the sea, too, but they stay in shallow water, where they feed on

▽ The whale shark is the largest of all fish. It grows to 49 feet in length and feeds, like the largest of the whales, by filtering vast numbers of tiny zooplankton from the seawater.

STUDYING THE OCEANS

Because humans cannot breathe in water, our knowledge of the ocean depths was first based on creatures caught in dredges or in nets dragged behind boats. Although these catches showed something of the wide variety of marine life, they did not include animals that lived attached to rock. Also, they provided little information about how the animals behaved. Fragile animals were broken up in the nets, and the deep sea was still largely unknown.

It was not until 1943, when Jacques Cousteau and Emile Gagnan developed SCUBA (Self-Contained Underwater Breathing Apparatus) that humans were able to swim to greater depths, breathing freely underwater, using tanks of compressed air. After this, the fantastic variety and beauty of the underwater world was made widely known through photographs and films. For the first time, scientists could directly observe and record marine animals living in their natural habitat, rather than in tanks.

Divers can only breathe air safely to a depth of around 197 feet. To explore deeper, small submarines called submersibles have been developed. These are designed to withstand the enormous pressure of the deep sea, and to take humans to great depths.

More detailed study is possible using a human-shaped suit, called JIM (right), which allows the person inside to walk on the seabed, take samples, and perform a variety of tasks.

This exploration of the deep sea has shown us that some of the most abundant animals in the sea are extremely fragile forms, too delicate to be collected by nets and trawls.

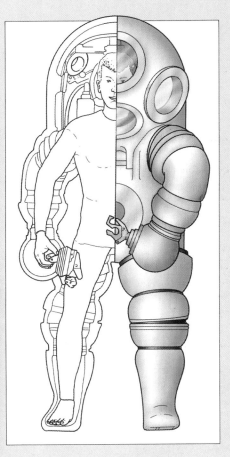

seaweed and sea grass. They occasionally come ashore, but cannot breathe easily without the support of water.

Seals, sea lions, and walruses all return to land to mate and to have their pups. Some seal pups can swim within hours of birth, but pups of other species stay on the beaches for weeks or months before going into the sea.

Few reptiles are truly marine, and because they are "cold-blooded," they are restricted to living in tropical seas. The brightly colored sea snakes of the Indian and Pacific oceans mate in the sea and give birth to one or two well-grown young, ready to swim off.

Turtles are the only other marine reptiles, but although they mate in the sea, they must return to land to lay their eggs in holes in sandy beaches. Many eggs perish and young turtles die mainly as a result of disturbance from humans and predation from birds.

It may seem strange to include birds under "strong swimmers," but several seabirds are indeed expert swimmers. Penguins have wings that are so well-designed for "flying" underwater that they cannot fly in the air. Auks (such as puffins and guillemots) can also swim well underwater but have retained the ability to fly.

▷ Sea snakes are the only reptiles alive today that are able to spend all their life in the sea. Their bright colors warn of deadly poison.

BETWEEN THE TIDES

The plants and animals that live between the high and low watermarks on the shore only have the benefits of living in the sea for part of the day. For the rest of the time, they have to cope with exposure to cold, heat, and drought.

Most true marine plants and animals cannot survive for long out of the water. Yet the animals and plants of the intertidal zone, between the highest and lowest points reached by the tides, have to live for lengthy periods out of the water. During this time, they have to cope with the extremes of hot and cold weather. They face a constant struggle to avoid drying out, but occasional soaking by fresh water in rainstorms can be equally damaging.

The plants and animals farthest down the shore are out of the water for a brief period only when the moon produces the extremes of high and low spring tide every 14 days. For most of the time, they have all the benefits of living in the sea, and as a result, are similar to the plants and animals that live deeper in the sea.

However, plants and animals at the top of the shore may be out of the water for 11 hours or more. They need to be highly adapted to live in this zone and deep-water species cannot survive there. Between these two extremes, the time that plants and animals spend out of the sea depends on how far up the shore they live. The result is a zonation of different species up the shore (see below and page 122).

The rest of this chapter describes the typical pattern of zonation in the North Atlantic. Tropical and subtropical shores also show a pattern of zonation, although the species are different.

black and yellow lichens

channeled wrack

high water mark

spiral wrack

bladder wrack

Channeled wrack

egg wrack

thong weed

coralline algae

serrated wrack

kelp

red a

low water mark

Egg wrack (left) and bladder wrack

△ A range of seaweeds grow at different points up a rocky shore, depending on how well they to survive out of the water. This produces a characteristic zonation. Animals show a similar zonation. Kelp only grows where it is always covered, except at the lowest tides. Egg wrack and bladder wrack, which grow on the middle shore, have bladders that hold them up toward the sunlight when the tide is in. Channeled wrack only grows at the very top of the shore and can dry out completely without ill effects.

Kelp

△ Flat periwinkles shelter among the mat of bladder wrack when the tide is out.

Life on the rocks

Although kelps grow anchored to rocks beneath the sea, their broad blades have no way of preventing water loss and they rapidly die when exposed to dry air. At the lowest spring tides, kelps may be briefly uncovered by the tides, but never for long enough to dry out. They therefore mark the bottom of the intertidal zone and cannot grow farther up the shore, except in rock pools. Sea urchins, sea anemones, sea squirts, and sponges shelter among the kelp, and are soon covered again by the advancing tides.

Above the low watermark, if the shore is not too battered by the waves, various species of wrack grow on the rocks. These brown seaweeds have flattened, leafy fronds, often with air-filled bladders along the fronds. These buoy them up toward the sunlit surface when the tide is in, to increase photosynthesis during their brief period underwater.

When the tide goes out, the wracks flatten against the rocks in a thick mat, so that only the top surface dries and the fronds underneath remain moist. Many shore animals shelter in this mat, including crabs and sea snails such as periwinkles and whelks.

Rock pools provide small oases of seawater in this rocky desert, allowing animals from deeper water to survive on the shore. In hot summer weather, however, the rock pools become dangerously warm and highly salty as water evaporates from their surface, while rain can make them harmfully dilute.

Limpets and barnacles live beneath the seaweeds and on more exposed, bare rocks, protected by their cone-shaped shells. When the tide is in, limpets move off to graze on algae on the rocks, and barnacles open the lid at the top of their shells to feed (see below).

All these animals are adapted to feed and even breed during the few hours when they are covered by the tide. When the tide is out, birds like oystercatchers and gulls search the shore for animals beneath the seaweed.

Several species of wrack grow at different levels up the shore, until, at the top of the shore, only channeled wrack can grow. This plant has grooved fronds, which curl up to reduce water loss during the long periods when it is left high and dry by the tide. Tiny periwinkles shelter among its fronds or in cracks in the rock.

Above this zone, the seaweeds give way to black, yellow, and gray lichens, which can survive the salt spray from waves splashing against the shore.

▽ Barnacles have been described as animals that stand on their heads and kick food into their mouths, using their feathery "feet."

STUDYING ROCKY SHORE ZONATION

Start at the foot of the tape and list in a notebook all the seaweeds and animals you recognize. Sketch any you do not know to check later in a book on seashore life. Repeat this process at intervals (perhaps every foot) up the tape. As you move up the shore, you will find that some species disappear and others appear.

When you get home, make a chart showing the length of the shore. Use the chart below as a guide. Draw a colored line for each species you saw, joining the lowest and the highest points at which they live on the shore. Rock pools will affect this pattern of zonation, so mark where they appear on your chart.

The best time to study the zonation of plants and animals on a rocky shore is during spring tides. These occur at the times of the full and new moon, which are shown in most encyclopedias. Tidetables from fishing shops or harbor offices will show the time of low tide, and you should start your survey just before then. Take great care, especially near deep water, and make sure you do not get cut off by the tide.

Stretch a canvas measuring tape in as straight a line as possible up the shore from the point where the sea is farthest out to where the grasses and flowers start. (Remember to retrieve the tape as the tide comes in.)

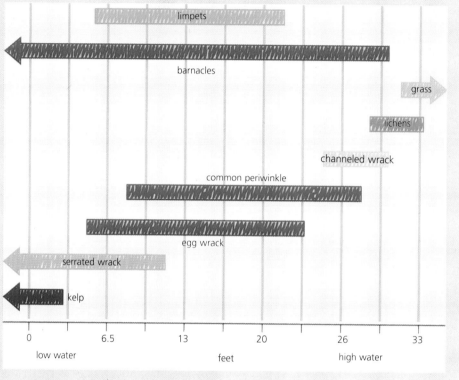

Sand and mud

Animals living on sandy or muddy shores can bury themselves to prevent drying out when exposed by the tide. However, water drains quickly from sand, so very little can survive in the upper parts of sandy shores, which are dry for longest. A few animals more typical of sandy seabeds live, well buried, on the lower shore.

Fine particles of mud are much better at holding water, and provide a safer refuge for animals when the tide is out. Many animals live in burrows in the mud, coming to the surface to feed when covered by the sea.

Various bivalve mollusks, including cockles and razor shells, worms such as rag worms and lugworms, small sand hoppers and crabs live buried in the mud.

Tiny spire-shell snails feed on algae on the surface. A few green seaweeds grow on the mud, especially where it is enriched with nitrates from the land. Various fish and prawns follow the incoming tide to feed on the mud life.

Bird-feeding stations

The biggest expanses of mud are found in sheltered estuaries, the broad mouths where rivers meet the sea. These can support incredibly rich life. In a square foot of mud, up to 1,000 rag worms, 42,000 spire-shells, and 63,000 mud-burrowing sand hoppers have been recorded (although not all together).

This rich food supply is very important for wading birds, which feed on mud creatures at low tide. However, estuary

shores rarely provide safe breeding sites and so the birds move inland or to the Arctic to breed in summer.

Estuaries, therefore, provide essential feeding stations for wading birds in winter and are vital stopping points during their migration. The numbers of birds at some sites can be impressive: 235,000 waders in Morecambe Bay in northeast England, 2 million in the Waddensea off the coast of Holland and Germany, and over 10 million in the Copper River delta in Alaska. Yet estuaries are often regarded by people as "waste ground" suitable for building factories, power plants, marinas, or dams for generating tidal power. Estuaries around the world are threatened by such developments, as well as by pollution and overfishing.

△ The mud of estuaries may seem empty at low tide, but it supports as great a weight of animals as the richest pasture. Huge numbers of wading birds, such as these western sandpipers (inset left), gather to feed on the abundant life. The deeper water of estuaries is also an important breeding ground for flatfish, such as flounder (inset right).

▷ The different sizes and shapes of wading birds' beaks allow them to feed in different ways in the mud, thus reducing competition for food at the surface. Short-billed plovers feed mostly on spire-shell snails on the surface. Birds such as the knot and redshank, with bills around 1 inch long, dig into the mud for small worms and sand hoppers, while curlews, with bills up to 4.7 inches long, dig much deeper for lugworms and deep-burrowing bivalve mollusks. The illustration also shows a lugworm living in its U-shaped burrow, taking in food down one tube and expelling waste out the other.

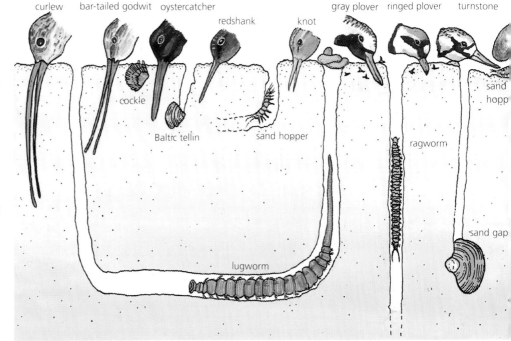

curlew   bar-tailed godwit   oystercatcher   redshank   knot   gray plover   ringed plover   turnstone

cockle

Baltic tellin   sand hopper

sand hopper

ragworm

lugworm

sand gap

LIFE ON THE COAST

The influence of the sea reaches many feet inland, creating a narrow strip of sand dune, salt marsh, or cliff around the land. Often humans have scarcely changed these coastlands, making them a valuable refuge for plants and animals.

△ Glasswort grows on muddy coasts around the world. Its cactuslike shape helps it to store water. It also gathers sodium salts from the sea. In the past, it was collected and burned, and glass or soap was made from its ash.

Anyone who lives near the coast knows how far the effect of the sea reaches inland. On stormy days, cars become white with salt and the salt spray is enough to kill delicate garden plants. But the sea also acts like a hot-water bottle, warming the land nearby.

The sea has another physical effect on the edge of the land. Waves pounding against exposed shores create towering cliffs. On more sheltered shores, the sea dumps sand and mud, building a new shoreline, which can, just as easily, be carried away by the next storm.

This constant battle between the sea and the land, combined with wind and salt spray, stops woodlands from growing on the coastal fringe, leaving open grassland. This grassland is often rich in plants and animals.

Drought in the marsh

At the top of muddy shores, a remarkable plant called glasswort grows abundantly. It looks like a miniature cactus, with a stubby stems. The glasswort stems trap mud from the tides, raising the level of the sea bottom and allowing grasses, rushes, and a variety of smaller, more colorful plants to begin to grow. These plants are covered only briefly by the highest tides, creating what is called a salt marsh. The plants bind the mud and prevent storms from washing it away.

Many salt-marsh plants have leathery leaves and fleshy stems, like desert plants. This similarity is no accident. Although the marsh is regularly soaked by the sea, salt-marsh plants have no way of extracting fresh water from the sea, just as we cannot drink seawater to quench our thirst. Their only source of fresh water is rain, so they must store this precious water as a desert plant does.

Furthermore, because the salt in the sea is so highly concentrated, it tends to draw water out of plant cells. This would cause most plants to wilt and die, but salt-marsh plants are protected by their leathery leaves and stems. Many also absorb salt from the sea so that the concentration in their cells matches the seawater, which prevents water loss. Excess salt is expelled by many salt-marsh plants out of special glands.

Once the mud is raised above the tides, rainfall turns the salt marsh into a fresh-water marsh. This is often covered by dense beds of reeds, which trap more mud washed in by the rain.

While the mud builds up inland, the glasswort at the seaward edge creeps outward. As a result, the salt marsh grows and any one spot on the shore will change over many years from open mud to low salt marsh, and eventually to rushy marshland – a change over time called a succession. If you walk inland across a salt marsh, you will see a similar change from bare mud to

◁ These barnacle geese are among the few animals that graze on salt marshes in winter.

hollows and pools between the dunes, and the combination of dry and damp soil encourages a rich mix of flowers. Lizards and snakes bask on the hot sand on sunny days, and gulls, terns, and larks nest among the vegetation.

Still further inland, as the influence of the sea declines, shrubs and trees can begin to grow, and the dunes give way to heathland or woodland – the final stage in the dune succession.

◁ Marram grass has tough leaves to survive in dry sand and deep roots to reach fresh water in the cold sand far below. Its roots spread out through the sand, binding it so that it builds up into towering sand dunes.

rushy marshland, but this occurs over distance rather than time, and is called a zonation. The growth of the salt marsh only stops when storms and currents wash the mud away as fast as it gathers.

Only a few geese and ducks graze on the tough and salty salt-marsh plants, and wading birds and ducks nest among the vegetation in summer.

Hills of sand

A similar succession happens on sandy shores, but with the important difference that sand grains can also be carried far inland by the wind. A few plants with tough, leathery leaves grow at the top of the beach. Windblown sand builds up around them, making low hummocks above tide level in which grasses and sedges can begin to grow. The most important of these on European coasts is marram grass, which grows up through the sand as the sand gathers around its stems, forming sand hills called dunes.

However, the cover of marram is not continuous. Patches of bare sand are left and this is readily blown away in the wind. As a result, the shape of the dunes keeps changing, like shifting sands in the desert.

Further inland, where the wind carries less sand, plants form a more continuous carpet, and the dunes are lower and more stable. Frogs and toads breed in damp

Rocky coasts

On cliffs and rocky coasts, there is a similar zonation of plants away from the sea, although this does not change over time in a succession. Above the lichens of the splash zone, a few plants with leathery and waxy leaves cling to the fast-draining rocks. Even on top of the cliff, wind and sea spray keep trees away, and a narrow band of grass or heathland forms, rich in flowers and butterflies.

▽ The Camargue is an area of sand dunes and salty lagoons where the River Rhône meets the Mediterranean Sea in France. It is an important home for huge numbers of water birds, including 10,000 or more greater flamingoes.

FORESTS IN THE SEA

Around 92,670 square miles of sheltered shores in the tropics – an area larger than the state of Florida – are covered with dense forests of mangroves. In fact, the name mangrove covers over 50 species of evergreen trees and shrubs that belong to several unrelated families, but all are adapted for growth in seashore mud. They have spreading roots to anchor them and small fingerlike roots that stick up through the mud, allowing the roots to "breathe." Most produce long, torpedo-shaped seeds that germinate in the mud, causing the forest to expand across the shore.

The tides regularly cover the base of these trees, and a wide variety of animals live in the trees' shelter. Barnacles cluster on the undersides of the lowest mangrove leaves. Fiddler crabs live in mud burrows and feed on dead animal matter among the roots. The large claw of the male fiddler crab is not used for eating, however, but for display, to attract a mate.

Mud skippers are fish that spend more time out of the water than in it. They move across the mud in a series of "skips" by wriggling their tails, and can even climb onto mangrove roots. When the tide is out, they bury themselves in the mud, with just their large eyes showing, watching out for danger.

Mangrove roots

Fiddler crabs

Mud skipper

THE HIGH-RISE LIFE

Puffin

Steep sea cliffs with broad ledges provide a safe nesting site for huge numbers of seabirds, but still allow easy access back to the sea. This is especially important when the young birds are ready to leave the nest and make their first, unsteady flight down to the sea. Each species selects a slightly different breeding site, as can be seen from this cliff in the eastern Atlantic.

Black guillemots (1) nest under boulders at the base of the cliff, just out of reach of the waves. Common guillemots (2) gather in great numbers on cliff ledges. They lay their eggs on the bare rock, rather than in a nest. The eggs are pointed at one end so that they roll around in a circle rather than rolling off the edge of the cliff.

Razorbills (3) prefer a ledge with a slight overhang, laying their eggs in a crevice or hollow in the rock. Shags (4), a type of cormorant, make an untidy nest of seaweed and twigs on broad ledges. Kittiwakes (5), a species of gull, build neat nests of moss, grass, and seaweed, which they "glue" onto jutting pieces of rock, using their droppings.

Gannets (6) make untidy nests of seaweed or grass on broad ledges or flat, grassy areas near the top of the cliff. Fulmars (7) have scattered nests on grassy ledges. Puffins (8) dig burrows or take over the burrows of rabbits on grassy slopes that hungry rats cannot reach.

Arctic skuas (9) attack puffins returning to their nests with food to make them drop the fish they are carrying. They then catch the fish in midair or snatch it from the sea surface. Arctic skuas nest on moorland some distance from the cliff.

ISLAND LIFE

Small islands, far from the nearest land, provide a safe nesting site for seabirds. Many of these islands are also inhabited by their own special species of land plants and animals, and thus tell us a great deal about the evolution of life on Earth.

St Kilda •

Galápagos Islands

Scattered around the oceans of the world are small islands. Some were formed when undersea volcanoes erupted to the surface. Others, in shallower seas, were formed by the growth of corals and became dry land when the sea level fell.

When the islands first formed, they must have been completely empty of life. Probably they were soon discovered by wandering seabirds, which still return today to rest and nest. But before long, the islands were colonized by other animals and plants, and today they support a rich variety of wildlife, much of it found nowhere else in the world.

The colonists

But how did plants and animals arrive on islands so far from the nearest land? Nobody can be absolutely sure of the answer, but a walk along a tropical beach gives a clue. Often, fruits are washed up on the shore from faraway places, and sometimes whole, living branches of trees are swept ashore. Some of these must have reached the shores of such isolated islands and begun to grow, bringing green life to the previously barren land. The seeds and spores of other plants – such as grasses, ferns, and mosses – can be carried long distances by the wind, and may have reached these islands and begun to grow.

Perhaps birds on migration were blown off course and landed on such islands. We know, for example, that small numbers of birds from North America are regularly blown across the Atlantic to Europe in stormy weather. The same could have happened to force birds to land on remote islands, although many more birds must have died without ever reaching land.

It is more difficult to explain how "stay-at-home" animals, such as snakes and lizards, could have reached these islands.

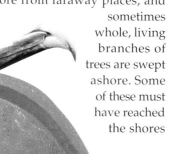

◁ Great frigate birds nest in bushes on islands throughout the Tropics where they are safe from hunters. The males inflate balloonlike scarlet throat pouches to attract females. Frigate birds feed on fish from near the surface of the sea and sometimes snatch food from other birds.

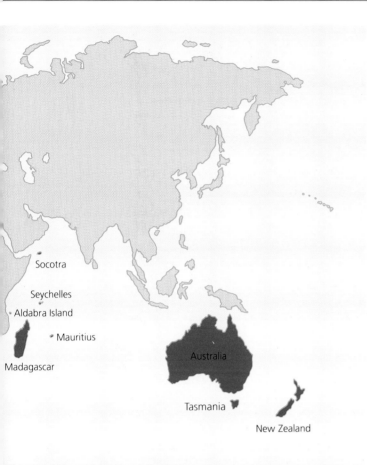

△ To survive on islands, animals must adapt to different ways of life. Marine iguanas on the Galápagos Islands dive beneath the sea to feed on seaweed. They hold their breath underwater and slow their heartbeat to save oxygen.

◁△ The sack-of-potato tree lives only on the island of Socotra, in the Indian Ocean. It has few leaves, but in winter it produces a mass of pink flowers, earning it the name "desert rose."

One possibility is that they were carried there as accidental passengers on the branches of trees that had washed ashore. Although the chances of their surviving on branches over such a great distance seem remote, only one pair – or even one pregnant female – would be needed to reach the island and begin breeding.

Certainly reptiles would have been able to survive such a voyage much better than mammals, as mammals need much more food just to stay alive. Perhaps that is why more lizards and snakes reached these remote islands than mammals.

When the colonists reached the island – however they got there – they would have found abundant food, and few other animals to compete with. There were no large hunters on the islands, because these could not survive a long ocean journey, so the newly arrived colonists had no enemies. In these ideal conditions, they must have bred rapidly.

With such a rapid increase in numbers, chance twists in the biological makeup of plants and animals are very likely. Such changes help drive the process of evolution. Life on the island would also have been very different from life in their original home, so the colonists would need different characteristics to survive. As a result, they began to change, over thousands of years, from the form taken by their ancestors back home, until new island species evolved.

◁ Madagascar, an island in the Indian Ocean, is the only home for lemurs, ancient relatives of the monkeys. The strangest of these is the aye-aye. It has large eyes to help it see at night and a long middle finger to poke for grubs beneath the bark of trees.

▷ The kakapo is a rare parrot from New Zealand that spends most of its life on the ground. It makes tracks through the forest at night as it feeds on ferns, mosses, and berries.

Island treasures

The animals living on ocean islands today often give a clue to their origins. Giant tortoises live on several islands, including Aldabra, the Seychelles, both in the Indian Ocean, and the Galápagos Island off the coast of South America. They are similar to tortoises found on the mainland, but are much bigger. Perhaps their large size helps them survive dry summers on the islands when little food or water is available.

Without enemies, some bird colonists lost the power of flight. Flightless rails – relatives of the coot – are found on several remote Atlantic islands, for example, and flightless penguins and cormorants live on the Galápagos Islands.

The most famous flightless bird was the dodo from Mauritius in the Indian Ocean. Despite its turkeylike size and shape, scientists think it evolved from a species of pigeon and became heavy and flightless because it had no enemies. However, its in-ability to fly meant it had little chance of escaping human hunters when they discovered the islands, and so the last dodo was killed in 1681.

Leftover islands

Some of the world's larger "islands" have a different history. About 300 million years ago, they were part of a huge super-continent, but they gradually became separated as slow movements in the Earth's crust caused the continents to drift apart. As the continents separated, they carried animals and plants with them. With fewer enemies on these smaller, isolated landmasses, some of these animals survived and evolved into the modern species we see today, although their relatives became extinct on the largest continents long ago.

In New Zealand, for example, a rare, lizard-like creature called the tuatara is the only surviving relative of an ancient group of reptiles. Fossils of an almost identical creature have been found from 200 million years ago. Other ancient animals, such as the flightless kiwi, live in New Zealand. However, birds also reached the islands on the wing later in their history, and have evolved there quite separately. So New Zealand also has a flightless rail, called the takahe, and a ground-living parrot called the kakapo.

◁▷ The marsupial mole (right) looks exactly like a European mole, but its pouch shows that it is a marsupial related to the kangaroo. Although doglike, the thylacine or Tasmanian wolf (left) was also a marsupial and carried its young in a pouch. The last living thylacine was seen in the 1930s.

CHARLES DARWIN AND HIS FINCHES

large tree finch

sharp-beaked finch

large ground finch

woodpecker finch

These four, similarly sized finches from the Galápagos Islands were among those that helped Charles Darwin form his ideas of evolution. The large ground finch has a heavy beak to crack open nuts, while the small ground finch uses its thinner beak to eat ticks off tortoises' backs. The sturdy beak of the woodpecker finch allows it to drill into wood for insects; sometimes it also uses a cactus spine to poke them out. The sharp-beaked ground finch has even been seen pecking at seabirds' feathers to drink their blood.

In 1835, the naturalist Charles Darwin visited the Galápagos Islands on a British research ship called HMS *Beagle*. What he saw there helped him form the theory that he later proposed in his famous book *On the Origin of Species*. According to this theory, animals could change – or evolve – over long periods of time as chance changes in their bodies increased their opportunities for survival in competition with other animals around them.

Among the animals that led Darwin to reach this conclusion were 13 species of finchlike birds on the islands. They were all very similar, sparrow-sized, blackish or brownish birds, but each species had a different-shaped bill. The only way Darwin could explain the appearance of the finches was to suggest that they had all originated from one seed-eating ancestor that had reached the islands. The various beak shapes had then evolved to allow the birds to take advantage of different foods, resulting eventually in new species.

The large island continent of Australia is home to an even more remarkable group of animals: marsupials. These give birth to tiny, hardly formed young, which they rear in a pouch of skin on the outside of their body. Marsupials first evolved when Australia was still joined to South America, but most South American marsupials were killed off by other, more successful mammals that invaded from North America much later. Today the only marsupials that exist in South America are 70 species of ratlike animals called opossums.

Isolated in Australia, however, the marsupials developed into many forms, often suited to the same way of life as more recently evolved animals elsewhere. Kangaroos, for example, are the Australian equivalent of antelopes, and there are marsupial squirrels, anteaters, moles, and cats. There was even a marsupial "wolf," the thylacine, although this is almost certainly now extinct even from its last home in wooded areas of Tasmania.

▽ It is not just tropical islands that have special wildlife. The St. Kilda wren is a subspecies of the common wren that is only found on the islands of St. Kilda, 118 miles west of mainland Scotland.

LIFE AROUND PEOPLE

Although humans have damaged and destroyed many of the world's wild places, they have also created new environments for some plants and animals. A variety of animals have even followed people and found homes in the middle of cities.

△ Woodworm, which can do a great deal of damage to timber in houses unless controlled by chemicals, are the young of the furniture beetle (above). This beetle once lived in the bark of old trees. Today they find damp timber and old furniture just as good to eat.

No other animal has made such an impact on the face of the Earth as humans. But some plants and animals have been able, over the course of history, to take advantage of the changes made by people, finding new homes and new sources of food as a result.

Scavengers must have followed early human hunters to take advantage of the leftovers from their kill, just as vultures benefit from a lion's kill in Africa today. As humans began to settle on the land and travel to different parts of the world, some animals followed them and made new homes, too.

Human followers

The first human farmers helped the house mouse move from its original home in the grasslands of Asia and the Mediterranean 5,000 years ago. Today it has spread to almost every part of the world. The black rat also came from Asia, where it lived in trees. Its climbing skills meant it could clamber up the rigging of sailing ships. It reached Europe in the 11th century and South America in the 16th century.

Unfortunately, the rats had fleas in their fur and these carried the bacterium that caused bubonic plague. One outbreak of the plague, called the Black Death, killed at least one-quarter of the people of Europe in the 14th century. Since then better hygiene, and the poisoning and trapping of the black rat has made it a rare animal in western Europe, found only in a few dock warehouses and on small islands.

The black rat's disappearance was helped by the spread of the brown rat from Asia across Europe in the 18th century. It reached England by 1770, and five years later it had crossed to America on ships, although it took another 80 years before the spread of railways helped it to reach the American West.

As humans set up house, other animals took advantage of the warmth and food. Beetles, moths, cockroaches, and even ants have moved around the world living in people's homes.

Rabbits were spread in a different way. They originally came from Spain, but the Romans and Normans moved them around Europe, keeping them in walled enclosures as a supply of fresh meat. Eventually, some escaped, and they soon built up to huge numbers in the wild. Later, humans took rabbits to South America, Australia, and New Zealand, where, without enemies, they have reached plague numbers.

◁ The house mouse has spread around the world with people, living mostly in houses and wherever food is kept or sold. It can survive outdoors, but rarely lasts long without unwitting help from humans.

◁ Chemical sprays and cleaner seed corn mean that colorful, weedy fields like this have become a rare sight in the United States and parts of Europe, although farming policies are helping their return.

Taming the wild

Humans have also "tamed" wild plants and animals for their own benefit. The first cereal crop grown by farmers in the Middle East about 9,000 years ago was the result of a natural cross between wild wheat and goat grass. Since then, careful breeding has ensured even larger heads of grain and bigger crops.

Other crop plants were taken to Europe from all around the world. Potatoes came from South America, tomatoes from Mexico, carrots from Afghanistan, cauliflowers from the Middle East, and rhubarb from China. Despite this, only 20 different crops provide most of the food humans eat today. Rice alone is the main food for almost half the world's population. Yet at least 75,000 species of plant are known to be edible. If we used more of these, perhaps we could feed the billion people who do not have enough food to eat.

Around 8,000 years ago, people in Asia also began to herd wild sheep and goats for the first time. Although these animals were wild, they were gradually tamed and became bigger and woollier as farmers chose the best animals from which to breed.

It took at least another thousand years before the ferocious wild ancestor of cattle, called the auroch, was first tamed. Modern farm cattle are much smaller than aurochs and have been taken all around the world, although the

Plant invaders

When humans began to clear the ground to grow crops in northern Europe, 5,000 years ago, plants such as nettles, docks and bindweeds quickly moved into the bare soil. Other plants were spread by farmers, mixed with their seed corn. The Romans, for example, took weeds such as corn marigold, corncockle and scarlet pimpernel with them to England, France, and Germany.

Later, when the Pilgrims crossed the Atlantic in the 17th century, they took plants such as plantain, dandelions, groundsel, and stinging nettles to America, where they soon became a problem as weeds. European settlers also

transported thistles and ragwort to New Zealand and Australia, where they are now troublesome weeds.

Today many of the old cornfield weeds have become rare because farmers use weed-killing chemicals, called herbicides. But some weeds are making a comeback, as some European policies set aside farmland to prevent too much wheat and barley from being grown.

▷ Humans have introduced animals to countries around the world for many different reasons. Cane toads, for example, were taken to Australia from South America in the hope that they would kill beetles that were damaging the sugarcane crop. Instead, they have become a serious nuisance, invading water tanks and swimming pools.

▷ The sheep kept by the first farmers were similar to these wild mouflon in the mountains of the Mediterranean islands of Sardinia and Corsica.

cattle of the Far East probably have a different wild ancestor.

City slickers

Perhaps the most surprising change is that some wild animals have made their homes in the middle of cities. The city pigeon, derived from the rock dove that nests on sea cliffs, is now common in cities from Rome to New York (where it was introduced by early settlers). Starlings roost in huge flocks on warm city buildings, coming from as far afield as Russia to spend the winter in London or Paris. Kestrels and even peregrine falcons have followed them there, nesting on tall buildings rather than on cliffs.

Mice and rats are common around cities, especially around abandoned buildings and places where food is sold. Raccoons and skunks also live in cities, raiding trash cans at night for food. Many other animals around the world have adapted to urban life, from lizards in the Tropics to wild herons in Australian cities.

▽ Raccoons are highly adaptable animals that originally lived in woods and scrubby areas in North America. Now they have moved into towns and cities, where they live in sheds and attics and raid trash cans.

IN AN ENGLISH CITY GARDEN

△ The flowerbeds in Dr. Owen's garden offer plenty of shelter for animals, and are full of nectar-rich flowers which attract insects like this elephant hawk moth (inset).

Gardens may seem dull compared to rain forests, but a naturalist called Jennifer Owen has shown how rich they can be. Over 11 years she studied the wildlife in her garden in the English Midlands. It was a typical middle-sized town garden with lawns, flowerbeds, vegetable patches, a few tall trees, and some overgrown bramble and nettle patches.

Dr. Owen counted the different plants and animals in the garden, and used traps to find animals she might otherwise have missed. Altogether she found an amazing total of over 1,700 species. These included 80 species of "weeds," 50 different birds, and 5 mammals. But the main variety was in the insects, including 323 species of butterfly and moth, 91 hover-fly species, and at least 553 species of parasitic wasp. Furthermore, at least 13 of these wasps had never been found in England before and two were new species never before recognized by scientists.

There is nothing particularly special about Dr. Owen's garden, although she does try to use "friendly" gardening methods without poisonous chemicals. Probably many other gardens, at least in temperate countries around the world, could produce as long a list if studied equally carefully, further proving how important gardens are for wildlife.

MAKE A WILD CORNER

△ Frogs and toads are often quick to find a garden pond like this in which to lay their eggs. If there are none in your area, you could bring in a little spawn or a few tadpoles from another pond, provided you leave plenty behind.

△ Buddleia is so attractive to peacock and other butterflies that it is called the butterfly bush.

Wherever you live, you can help your local wildlife. The easiest way to start is to feed the birds (on a windowsill if you have no garden). Peanuts and packaged wild-bird food are much better for them than bread, but scraps such as bacon rind, stale cheese, and rotten fruit will also be appreciated. It is probably better to stop feeding in summer when wild food is plentiful, and you should always make sure you feed the birds well away from cats. Remember that in icy weather, unfrozen water can be even more important than food.

Better still, if you have a garden, plant trees to feed the birds. Choose species that belong in your area and produce lots of fruits and seeds. In the United States, mountain ash, hawthorn, birch, and cherries are especially good. Thistles and a plant called teasel provide lots of seeds for small finches.

To attract butterflies, plant flowers that produce plenty of nectar, such as thyme, knapweed, Michaelmas daisy, buddleia, and ice plant. Butterflies also need somewhere for their caterpillars to live. Stinging nettles, brambles, violets, grasses, bird's-foot trefoil, garlic mustard, and rosebay willowherb are among the plants on which butterflies choose to lay their eggs.

A pond, even if made from an old bath or sink, will soon be taken over by lots of water animals. Other animals will welcome a pile of old logs as a place to shelter.

THREATS TO LIFE

The balance of nature is delicate but essential
for life. Humans have upset that balance,
stripping the land of its green cover, choking the
air, and poisoning the seas. The future for living
things on Earth is uncertain, unless we humans can
learn the lessons of ecology and begin to respect
our planet.

DISAPPEARING WORLD

Many animals and plants have become extinct because they have been hunted and collected or their homes have been destroyed. Numerous others are now desperately rare. Efforts to conserve the rarest species have started in many countries, but the destruction of wildlife and wild places continues.

Some species of plant and animal die out naturally because more recently evolved species are more successful at competing for food and living space. Others have become extinct because of changes in the planet or because of natural disasters. Dinosaurs, for example, probably disappeared because the climate became cooler, perhaps when a giant meteorite collided with the Earth, throwing up a cloud of dust that blocked out the sun.

Today, however, many more species are in danger of becoming extinct than can be explained by these natural processes. The difference is entirely due to the excesses of humans, hunting and collecting animals and plants or destroying their habitats.

Gone forever

Whereas wild-animal hunters live in balance with their prey, we humans often hunt far more than we need to. If we go on killing animals faster than they can breed, they are bound to die out eventually.

The story of the passenger pigeon (see right) shows how even common animals can be made extinct by a combination of hunting and habitat destruction. Many other animals have met the same fate. The dodo, for example, was hunted for its meat, and its eggs were eaten by rats and dogs, which people had taken accidentally to the island of Mauritius where it lived. The dodo could not cope with so much killing, and by 1681 it was extinct.

△ The African violet is a common house plant. Yet it is one of the 20 rarest wild plants. It is found only in a few forests in Tanzania, where they are being rapidly destroyed by local people.

There never were many dodoes on the small island of Mauritius, but the quagga, a kind of zebra, once roamed the plains of southern Africa in huge herds. However, when white settlers reached this part of the world, they began to slaughter these animals for their meat and hides. By 1883, the quagga, too, was extinct.

There are other well-known examples, like the great auk, the Carolina parakeet of the United States, and the bluebuck in Africa. Other extinct species are almost unknown, such as the indefatigable rice rat and the mysterious starling, but each extinction represents the tragic loss of something irreplaceable.

In fact, only around 95 species of bird and 40 species of mammal have become extinct in the last 350 years. Today, however, seven times that number face extinction in the near future. If invertebrate animals and plants are included, the total number of species in danger is 20,000. But the main threat today is habitat loss rather than hunting.

◁ So many quaggas were killed by white settlers in South Africa that they became extinct in the wild during the last century. A few, like this one, lived on in zoos, but the last quagga died in Amsterdam in 1883.

THE PASSING OF A PIGEON

There is no animal that better proves the greed and destructiveness of humans than the passenger pigeon of North America. These handsome birds were once among the most numerous animals on Earth.

In 1810, one of their nesting colonies covered over 116 square miles and held around 30 million birds. Trees were broken by the weight of nesting birds and wild pigs fed beneath the trees on dead chicks. On one occasion in Kentucky, a naturalist watched a huge flock fly past him for over four hours. He calculated that there were 2,230 million birds in the flock.

Such huge flocks tempted local people to shoot them in large numbers. Soon pigeon hunting became a business. By 1870, 10 million dead pigeons a year were being sold in cities throughout the United States at a few cents each.

At the same time that this slaughter was taking place, the forests in which the pigeons lived were being felled to make way for the growing human population in the United States. This made it difficult for the pigeons to find enough acorns and berries to eat or suitable places to nest. As a result, they could not breed properly.

By 1890, only small flocks were left, but still the shooting continued. In 1907, the last wild passenger pigeon was seen near Quebec – and shot! A few birds had clung to life in zoos, but they did not breed well and died one by one. The last survivor died in the Cincinnati Zoo in 1914, barely a century after the huge flock filled the sky over Kentucky.

△ In 1813, the artist John James Audubon, who painted the passenger pigeons above, saw a huge flock of the birds that "obscured the noon-day light as by an eclipse."

The tragedy of the forest

The destruction of the rain forest is the most urgent problem, partly because at least half the world's species live there. In recent decades, humans have felled more and more forest, both for timber and to clear the land beneath for agriculture.

As a result, only half the world's rain forest remains, and the rest is disappearing fast. An area twice the size of Maine is cleared every year, which is equivalent to a football field every three seconds!

▷ Half the world's 4 billion acres of rain forest has already been destroyed, and the rest could go within 40 years if the rate of felling continues.

◁ The destruction of rain forests has effects that reach far beyond the forest. With the trees gone, the thin forest soil is washed down rivers into the sea. There it clogs up and destroys coastal coral reefs, killing the wildlife that lives on them and threatening important breeding areas for valuable fish.

As the forest is destroyed, so also will the plants and animals of the forest disappear with it. Yet the tragedy of the destruction of the rain forest is not just the wildlife that is being destroyed, but the valuable resource that humans are losing.

▽ Native peoples, like the Yanomami of the Amazon, still live a simple life in the rain forest, hunting, gathering edible plants, and growing a few native crops. Like the animals, their future depends on the forest not being destroyed. Yet in 1993, 40 Yanomami were massacred by gold prospectors who did not want the rights of the Yanomami standing in the way of their own greed for gold.

And the same is true for all other habitats that are also being destroyed.

Felling the rain forest causes many problems for people. The forest trees act like sponges, soaking up water during storms and slowly releasing it into the atmosphere through their leaves. With the forest gone, the rain now pours down rivers, causing devastating floods. Without the trees to release water back into the air, the floods are followed by equally devastating droughts. In combination, floods and droughts have killed thousands of people in tropical countries in recent years as a direct result of forest clearance.

Yet the soil left after the trees are cleared is much too poor to support farm crops or grazing animals for long. Furthermore, without the trees, the soil is soon washed away by the rains, leaving behind a "wet desert" of little use to man or beast.

With the loss of the forests, humans also lose the potential benefits that could come from rain forest plants. Scientists estimate that 2,000 of these plants might have value in treating cancer, for example, and others might be useful against AIDS. To date, fewer than one-hundredth of all rain forest plants have been tested for their value to medicine, yet thousands of them could be extinct within 20 years unless the forests can be saved.

So humans will suffer in the future if we allow the rain forest, the richest habitat on earth, to be destroyed for short-term greed. And the same is true for all the other habitats that are being destroyed.

Protecting the rare

Early in this century, a few people became alarmed at what was happening and the first societies were set up to protect, or conserve, wildlife. Many other conservation organizations fol-lowed, including the most famous of all, the World Wildlife Fund, now called the World Wide Fund for Nature (WWF).

Through their efforts and those of the many people who supported them, governments were persuaded to do more to conserve wildlife. International laws and agreements were set up to stop wildlife and wild places from being destroyed, and many projects were begun to conserve endangered species. There were several conservation successes, including the Arabian oryx and California sea otter, but other attempts were much less successful.

Until recently, Project Tiger was thought to be a great success story. Great sums of money were spent setting up reserves for the tiger in India and elsewhere. These reserves certainly helped,

▷ The California sea otter was thought to be extinct until a small group was discovered on a remote coast in 1938. With careful protection, numbers have now built up to over 2,000, but a single oil spill could wipe out the entire population.

but today the tiger is more at risk than ever, with only about 4,700 left and numbers falling fast. Poachers can get up to $10,000 for a tiger's skin or its bones, which are powdered and sold in China as "medicines" to protect people against everything from malaria to evil spirits.

Despite worldwide efforts, giant pandas, mountain gorillas, African elephants, and rhinos continue to be in great danger. Laws to protect them are not strong enough when there are starving people desperate to make money from hunting them.

▽ Rhinos are still hunted for their horns, which are used in some parts of the world to make "medicines" and handles for daggers. The black rhinoceros was once the most common species, but hunting has reduced its numbers from 65,000 in 1970 to 2,400 in 1992.

WHALES TO THE SLAUGHTER

In the 50 years up to 1975, 1.5 million whales were killed for their meat, blubber, and bones. So many were killed that the blue whale – the largest animal that has ever lived – was brought to the verge of extinction. The whalers then moved on to smaller species: fin whales, sei whales, and finally even small minkes, until they too became rare.

In 1982, the International Whaling Commission – an organization of whaling and nonwhaling countries – was forced into voting for a 10-year ban on all commercial whaling. Some member countries of the commission, including Japan, Norway, and Iceland, did not accept the ban, although they all stopped hunting whales for money. However, the Japanese killed over 300 whales a year for "scientific research" and sold the whales afterward!

The partial ban helped the numbers of whales to build up a little. But then the Norwegian government announced that it thought the numbers of minke whales had increased so much that full-scale commercial whaling could safely start again. They described minkes as "the rats of the sea," saying they killed large numbers of valuable fish. They also argued that Norwegians had a traditional right to hunt whales.

In 1993, the Norwegians killed 226 minke whales in the North Atlantic Ocean (69 of them for "scientific research"). Some of the harpooned whales took up to 30 minutes to die. Japan is also planning to resume whaling. Many scientists think the future for whales is again in doubt.

△ The last wild Arabian oryx was probably killed in 1972, but careful breeding in zoos has allowed for some oryx to be taken back to the wild. A herd of over 100 animals now roams the deserts of Oman.

Life-support systems

△ Although sharks, like these hammerheads, are still common, they too are under threat. An American doctor has claimed that their bonelike cartilage is a cure for cancer. As a result, Pacific islanders can earn $5,000 for every shark they kill.

Increasingly, conservation efforts are being targeted at protecting wild places, since these are the essential habitats for wildlife. Governments around the world are beginning to appreciate that these wild habitats are a life-support system for the planet. They also recognize that many wild plants and animals might offer benefits to people in the future, providing food and medicines, the opportunity for scientific study, and the pleasure of just watching and enjoying them – but only if they still survive.

In 1992, at an international meeting in Rio de Janeiro, Brazil, many of these governments agreed to an international treaty to protect what is called biodiversity (the complete range of plant and animal species and the many values locked up within them). But few governments have yet made real efforts to live up to that promise, and the desire for short-term profit is still seen as more important than the long-term need to protect species.

The biggest threat of all is the sheer number of people in the world, all strug-gling to find enough space in which to live. The human population today is around 5.3 billion, but it is expected to double in the next 40 years. By then, the only space for wildlife might be in zoos and safari parks.

△ Spix's macaw is as rare as an animal can be! Only one male is known to survive, living in a woodland in Brazil where this photograph was taken in 1990. About 30 of these birds live in captivity, and there are plans to release some of the females so that they might breed with the lone male. Although the macaw is of no financial worth to humans, its extinction would add to the loss of biodiversity, which the world's governments have pledged to protect.

WASTING THE WORLD

Industries produce vast quantities of
waste chemicals. Some of these are dumped onto the land,
and others are released into water or the atmosphere.
A few are directly poisonous to wildlife, but
others are causing long-term changes that
could threaten all life on Earth.

Factories, power plants, and automobiles bring humans many benefits, but they can also cause many problems. We are using up energy supplies and raw materials at an alarming rate, and at the same time we are producing vast quantities of waste products that we cannot cope with.

For a long time, the land, sea, and air were regarded as convenient places to dump these wastes. Now people are beginning to realize that this is creating major problems for the future.

In the United States alone, every household produces at least a ton of trash every year. Although a great deal of this could be recycled, most of it is dumped on the land. Industry produces even more waste, so large areas of valuable countryside are used up as garbage dumps.

Chemical cocktails

Poisonous chemicals in these wastes are often washed into streams, poisoning the life there. Many countries have been left with thousands of toxic waste dumps, which will cost enormous sums of money to clean up. But if these deadly wastes are not dumped on land, they have to be

△ In the 1960s, many peregrine falcons in the United States and Europe failed to breed, or they laid thin-shelled eggs that cracked. These problems were caused by pesticides in their food that accumulated in their bodies, affecting their reproductive systems.

burned or shipped to developing countries, which results in equally damaging effects.

Chemicals get into the natural environment in other ways, too. Huge quantities of chemical fertilizers have helped increase crop production to feed the growing world population. Unfortunately, however, a large proportion of the chemicals is washed into

◁ Gulls may benefit from human rubbish tips, but in the process they can catch food poisoning and spread it to humans when they roost on water reservoirs.

POISON FROM THE SKY

Since the last century, smoke from factories has hung over towns in unpleasant smoky fogs called smogs. One particularly severe smog in London in 1952 caused the death of 4,000 people, mostly from lung diseases.

The response to these smogs was to try to clean up the air. Taller factory chimneys were built to take waste gases higher into the atmosphere and new power stations were situated well away from cities. This policy made cities cleaner, but only by spreading the problem further afield. Now, Scotland and Scandinavia suffer from the air pollution of northern England, and Canada from the waste gases produced in the northern United States.

The most damaging pollutants are the invisible gases produced by factories, power plants, and cars. Sulphur dioxide from burning coal and oil, and nitrogen oxides from power plants and car exhausts are carried great distances in the wind. They dissolve in water vapor in the atmosphere to form acids. These then fall to the ground in the form of rain, which in some places can be more acidic than lemon juice. The term "acid rain" is often used for this, but snow, clouds, and fog are even more damaging.

Trees are particularly badly affected because they gather the acid from clouds or mist that hangs for long periods around their leafy tops. Over half the forests of western Germany, Norway, and parts of eastern Europe have been severely damaged by acid and other pollution.

The acidic water runs into streams and lakes, dissolving aluminum from the soil as it goes. The water becomes so acid that it kills insect larvae in the streams, and the aluminum clogs up the gills of fish, causing them to suffocate. Thousands of lakes in the eastern United States are now so acid that no fish can live in them.

The polluting gases from factories can be controlled by using chemical cleaners, and catalytic converters reduce the wastes from car exhausts. But these are expensive. In the long term, only careful use of energy, cleaner factories, and fewer automobiles will reduce acid pollution.

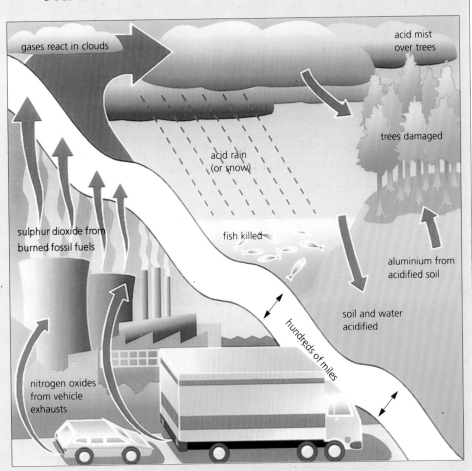

△ The origins of acid pollution

▽ Trees dying from acid pollution

◁ Two-thirds of the sewage from Mediterranean countries is pumped into the sea without treatment, encouraging deadly growths of bacteria and algae, and killing marine life.

▽ Until recently, a chemical called TBT was added to paints for yachts to prevent barnacles from settling on their hulls. But when the TBT dissolved in seawater, it made female dog whelks (inset) turn into males which could not breed.

lakes and streams, where it knocks the natural system out of balance, killing many water creatures.

Poisonous chemicals are sprayed onto fields to kill pests or to control weeds. As with fertilizers, these pesticides have brought many benefits, helping to increase crop yields and control dangerous diseases spread by insects. But they also cause untold damage to the natural world.

Pesticides like DDT and dieldrin remain powerful for a long time. They are washed into the soil and water, where they build up in the bodies of plant-eating animals and are passed on to the animals that hunt them. This continues up food chains until so much builds up in the top predators that they die.

When the effects of these long-lasting pesticides were recognized, they were banned in the United States and Europe, but they continue to be used in some developing countries.

Polluting the seas

The oceans are the "sink" of the planet. Anything that washes into streams or settles from the atmosphere ends up in the sea. The effects of so much pollution on marine life are unknown, but they are probably severe and worldwide. DDT, for example, has been found in seabirds in the Antarctic and in fish 9,800 feet down in the ocean depths.

Vast quantities of oil contaminate the sea. About 408,000 tons a year come from crashed oil tankers, such as the *Exxon Valdez* (see opposite), but over three times as much comes from the routine cleaning out of tankers at sea (about 727,000 tons a year) and from wastes washed down street sewers and other urban wastes (another 727,000 tons). Although the oil is eventually broken down by bacteria, too much in one place can kill large numbers of marine plants and animals.

Huge quantities of sewage and chemical wastes are also pumped into the sea. For example, an extra 771,000 tons of nitrogen chemicals are put into the North Sea alone every year as a result of human activities. In some cases, this has caused countless deaths of fish and seabirds, but there are also likely to be long-term effects as these wastes build up in the sea.

ALASKAN DISASTER

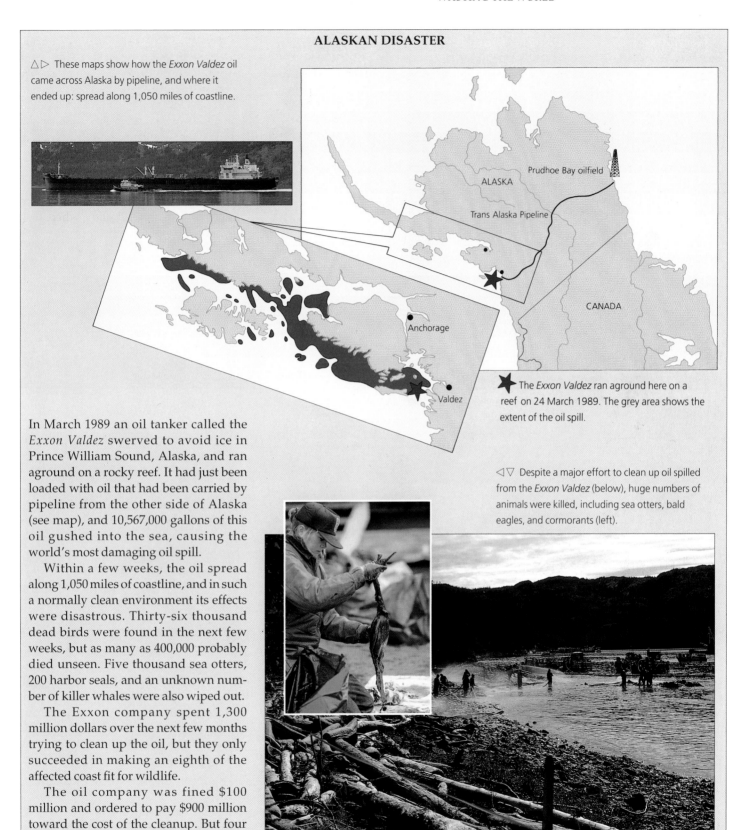

△▷ These maps show how the *Exxon Valdez* oil came across Alaska by pipeline, and where it ended up: spread along 1,050 miles of coastline.

⭐ The *Exxon Valdez* ran aground here on a reef on 24 March 1989. The grey area shows the extent of the oil spill.

◁▽ Despite a major effort to clean up oil spilled from the *Exxon Valdez* (below), huge numbers of animals were killed, including sea otters, bald eagles, and cormorants (left).

In March 1989 an oil tanker called the *Exxon Valdez* swerved to avoid ice in Prince William Sound, Alaska, and ran aground on a rocky reef. It had just been loaded with oil that had been carried by pipeline from the other side of Alaska (see map), and 10,567,000 gallons of this oil gushed into the sea, causing the world's most damaging oil spill.

Within a few weeks, the oil spread along 1,050 miles of coastline, and in such a normally clean environment its effects were disastrous. Thirty-six thousand dead birds were found in the next few weeks, but as many as 400,000 probably died unseen. Five thousand sea otters, 200 harbor seals, and an unknown number of killer whales were also wiped out.

The Exxon company spent 1,300 million dollars over the next few months trying to clean up the oil, but they only succeeded in making an eighth of the affected coast fit for wildlife.

The oil company was fined $100 million and ordered to pay $900 million toward the cost of the cleanup. But four years later, sea otters were still dying from the long-term effects of the oil. Colonies of guillemots and ducks in the area were failing to breed, and herring were being born with twisted spines,

distorted fins, or no jaws as a result of chemicals from the spill.

After the spill, a local Alaskan community leader summed up his feelings:

"Never in all our tradition have we thought it possible for the water to die. But it is true. We walk our beaches and, instead of gathering life, we gather death."

LIFE IN THE GREENHOUSE

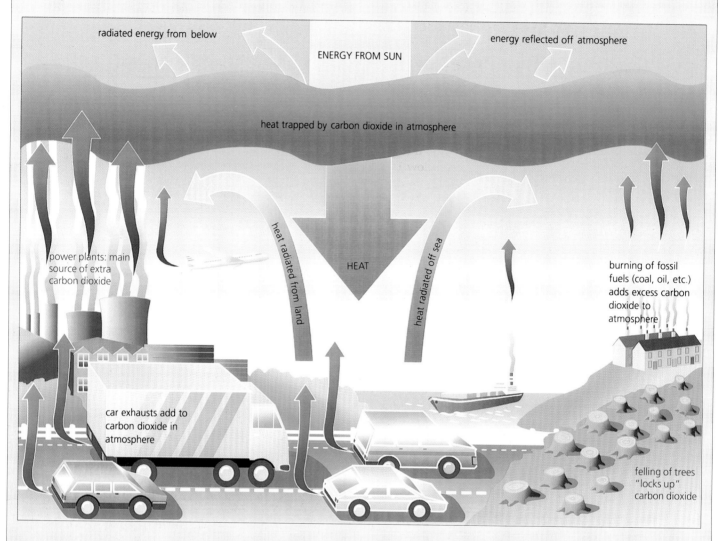

radiated energy from below

ENERGY FROM SUN

energy reflected off atmosphere

heat trapped by carbon dioxide in atmosphere

HEAT

heat radiated from land

heat radiated off sea

power plants: main source of extra carbon dioxide

car exhausts add to carbon dioxide in atmosphere

burning of fossil fuels (coal, oil, etc.) adds excess carbon dioxide to atmosphere

felling of trees "locks up" carbon dioxide

△ How the greenhouse effect works

The Earth is like a giant greenhouse in space, but, instead of glass, some of the gases in the atmosphere help to keep the planet warm. These gases let heat from the sun reach the ground, then trap it there, helping to keep temperatures in the ideal range for life.

Unfortunately, however, human activities are upsetting the natural balance of gases in the atmosphere. Excessive amounts of carbon dioxide from the burning of coal, oil, and gas are building up in the atmosphere, and trapping too much heat.

Forest clearance makes this problem even worse. Living trees use up carbon dioxide in photosynthesis as they grow, but carbon dioxide is released back into the atmosphere when they are cut down or burned.

Other gases produced by humans add to the greenhouse effect, including CFCs (see opposite). Scientists think that the continuing buildup of all these gases in the atmosphere could increase the average world temperature by 37°F by 2070. That would be the warmest the world has been in 2 million years.

An increase in temperature may seem welcome in cold climates, but the overall effect is very difficult to predict. It could mean more storms, droughts, and floods, along with all the damage these natural disasters inflict.

The warming would be greatest at the Poles. This would cause the ice there to melt, so that sea levels would rise by at least three feet, flooding many low-lying coasts and estuaries. Many tropical coral islands would disappear under the waves.

The changes would happen so quickly that plants and animals would be unable to cope. Nobody can predict the effect on the world's wild habitats. Planting trees to soak up carbon dioxide would help, but it would take an area of new forest the size of Argentina to cope with all the cars in the world, not to mention the waste gas from factories and power plants.

The only way to stop these changes is to control all greenhouse gases and, above all, to greatly reduce our use of energy and thus slow down carbon dioxide output.

The choking air

Although the consequences of land and sea pollution are often easy to see, the effects of air pollution are just beginning to be understood. In the past, smoke and invisible gases from factories were a major concern in cities, damaging buildings and causing serious health problems for humans.

Today, smokeless zones control the smoke, and tall chimneys pump pollution high into the atmosphere, away from the streets below. Unfortunately, however, that just shifts the problem elsewhere. The main concern now is pollution in the atmosphere, which crosses national borders and thus is much more difficult to control.

The atmosphere is the engine that drives the climate upon which all life depends. Anything that affects the atmosphere could therefore have serious effects on all wildlife.

Accidents such as those in Bhopal and Chernobyl show how deadly and widespread the damage from air pollution can be. But the most serious long-term effects are due to gaseous wastes from factories and power plants. These are causing acid rain, a change in the world's climates, and the destruction of the Earth's protective ozone layer. Unless controlled, such developments could eventually make the planet uninhabitable.

△ In 1984, a poisonous gas leaked from a pesticide factory in Bhopal, India. It killed 2,500 people, and more than 25,000 people were blinded or suffered other injuries.

△ When the Chernobyl nuclear power plant in the Ukraine exploded in 1986, radioactive wastes were carried to England and Scandinavia, where thousands of sheep and reindeer had to be slaughtered because their meat was unsafe. About 20,000 people across Europe may die from cancers caused by the radioactivity. The picture shows the beginning of the cleanup operation in the Ukraine.

A HOLE IN THE SKY

High in the atmosphere around the Earth is a band of ozone gas, which is constantly maintained by lightning and ultraviolet radiation. At ground level, ozone is poisonous, but in the atmosphere it acts as a shield, protecting the Earth from harmful ultraviolet rays.

In 1982, scientists in Antarctica discovered that the ozone layer above them was getting thinner. The cause is a buildup in the atmosphere of chemicals called chlorofluorocarbons (CFCs). These are used in aerosol spray cans, foam plastics, refrigerators, and air-conditioning systems. When these products are dumped, the CFCs slowly rise into the atmosphere, and, in the Antarctic winter, they go through complex chemical reactions that destroy the ozone.

With the ozone shield damaged, more ultraviolet radiation can reach the

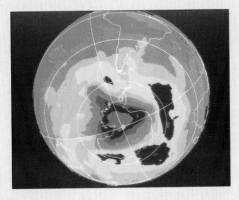

The ozone hole over Antarctica in November 1992

ground. This can lead to an increase in the occurrence of eye disease and skin cancer in humans. It could also damage the proteins and DNA in the cells of humans and wild plants and animals.

There are already signs that increased ultraviolet radiation may be killing some of the tiny plants (phytoplankton) that are the basis of all food chains in the Antarctic Ocean. Nobody knows what long-term effect that might have.

The ozone "hole" appeared first over the Antarctic because of the severe cold there, but since then a similar hole was found in springtime over the Arctic. This reached south to parts of Russia, Scandinavia, and Scotland, where many people live.

Because governments are worried about the effects of ozone thinning on human health and the environment, they have agreed to halve their use of CFCs by the end of the century. The problem is that CFCs remain in the atmosphere for 100 years. Even if we stopped using them tomorrow, they would build up in the atmosphere for another 10 years and the ozone hole would continue to worsen.

LESSONS FOR LIFE

An understanding of the natural world
can teach humans many lessons about living in balance
with the environment and leaving the planet
in a fit state for the future.

The science of ecology helps us realize that our small planet is not inhabited by a jumble of species living side by side. All plants and animals – including humans – have evolved to exist together in a finely tuned balance. But that balance is very easily upset.

More and more people are beginning to understand that message. They are trying to adjust their lives to be "friendly" to the environment, and they want governments to do the same. The result has been the growth of what is called "green" politics, with people doing whatever they can to

▽ Gases from automobiles, factories, and power plants have reacted in sunlight to form this eye-stinging smog over Los Angeles. Car exhausts can be made cleaner, but the best solution is for people to use cars less.

△ These members of Greenpeace, a conservation organization, are trying to get between a Japanese whaling boat and the whale it is hunting, to protect the animal and draw world attention to the continuing slaughter of whales.

stop damaging activities, protect natural sites, and persuade governments to do more for the environment. As a result, politicians around the world claim they have "gone green" and adopted environmental policies.

Mostly, though, their ideas are limited and short-term. They are concerned with protecting human health and making towns nicer places in which to live. They have done little to tackle the major problems affecting our planet, such as the greenhouse effect or the spread of deserts.

Changing ways

If humans are to survive, we need to protect the natural wealth that makes our life-style possible. We must look into the future, not just for a few years but for thousands of years, and use the knowledge of ecology to ensure that our actions are not damaging.

That will mean great changes in the way we live, especially in industrial countries. We will need to recycle more natural resources and waste far less. To reduce pollution and protect limited supplies of coal and oil, we will need to use cleaner sources of energy, conserve energy more, and question whether we really need to use so much energy.

HOW TO SAVE THE WORLD – OR AT LEAST MAKE A START!

Although only governments can make many of the changes needed in the world, there is plenty that young people can do.

SPEAK OUT
If there are things happening that you dislike, write letters to the people or governments concerned. Governments do listen to what young people say, and are often particularly impressed by letters from a foreign country.

WATCH OUT
If you see damage being done in your area, write to the local council, conservation organization, or newspapers.

JOIN UP
There are many clubs for young people, locally and nationally. They will tell you what is happening in the world and how you can help.

PAY UP
Conservation charities always need more money for their work. Save some of your own money for them or organize a project or an event to raise money.

CLEAN UP
Although insurance requirements usually mean you cannot do organized conservation work until you are 16, you could help with a local trash cleanup or make a wildlife corner in your garden or at school.

RECYCLE
Collect glass, tin and aluminum cans, and waste paper for recycling, to save raw materials, and to reduce dumping.

THINK IT OUT
Think carefully about everything you buy. Is it really necessary? Is there a more environmentally friendly alternative? You can also help family and friends make these decisions.

FIND OUT
Perhaps most important of all, learn all you can about the wildlife around you. The more we all know about the natural world, the more we can do to protect it.

▽ When world governments met for an Earth Summit in 1992, a million people made a pledge for the future of the planet. This pledge, in the shape of a leaf, was added to this Tree of Life – people-power at its best.

PLANET GAIA

△ Algae in the sea may control the whiteness of clouds and thus help to keep the Earth comfortably warm. But if the seas are polluted and the algae damaged, this temperature control could be destroyed.

Scientists are discovering that plants and animals have a vital role in maintaining the environment they need in order to survive. In photosynthesis, for example, plants absorb carbon dioxide and give out oxygen, which animals need in order to breathe.

Just by living, therefore, plants and animals control the balance of gases in the atmosphere on which life depends. These gases also allow the natural greenhouse effect to keep temperatures in the range needed by living things.

Living things have other effects on the environment. Transpiration from leaves (see page 51) maintains moisture in the air, while dead plants help make the soil. Nutrients are constantly stirred up by soil animals and spread over the soil's surface by animals in their droppings.

Just as a household thermostat switches on a heater when it gets too cold, these natural processes allow small balancing actions that maintain the conditions for life on Earth. Scientists think there might be an even more remarkable control mechanism. They believe that tiny algae in the sea may help to regulate temperatures at sea and on land. The algae produce a gas called dimethyl sulphide, which rises into the atmosphere and reacts with sunlight to form tiny particles. These particles absorb moisture, creating small water droplets. When there are many of these water droplets in clouds, the clouds become whiter and reflect more sunlight back into space.

Scientists think this link between algae and clouds might act as a sensitive temperature regulator for the planet. If lots of water droplets form in the atmosphere, more sunlight will be reflected and temperatures at the surface will fall slightly. But at this lower temperature, the algae produce less dimethyl sulphide, so fewer water droplets form in the atmosphere. This allows more sunlight through the clouds, raising the sea temperature and restoring the balance.

△ If we cut off a finger, without a doctor's help we would soon bleed to death. When we fell huge areas of the rain forest and do not repair the damage, the Gaia Hypothesis suggests that the planet also will "bleed to death," as a result of climate effects, pollution, and the loss of the natural wealth of species.

Temperatures on Earth might, therefore, be controlled in almost the same way that a lizard can warm itself up by basking in the sun or cool itself down in the shade. Furthermore, water transports nourishment around the planet in the same way as blood transports nourishment in the lizard's body.

The planet, therefore, behaves almost as if it is a living thing. The scientist James Lovelock has called this "living planet"Gaia, after the Greek goddess Mother Earth.

This "Gaia Hypothesis" is not meant to be a complete truth. But it does help us think about how we should treat the Earth. We are part of Gaia, like all animals and plants. The Gaia Hypothesis reminds us that if we damage the planet, we also damage ourselves.

We will need to change our ways of life to make travel less essential. When fixing the price for products, we will need to add on the true cost of cleaning up the pollution caused in their manufacture and disposing of them safely afterward.

We will need to use natural fertilizers instead of energy-expensive chemicals. We will need to replace poisons with natural ways of controlling pests, and remember that they only become pests because we create unnatural conditions in which they can flourish.

Most important of all, we will need to consider how many human beings the planet can support. The population is increasing fastest in the poorest countries, where a large family means security for the future. We will only limit population growth if we overcome the problems of poverty and unjust working conditions that make people want a large family.

△ Voles can sometimes cause serious damage to young conifer trees. However, rather than using dangerous chemicals to poison the voles (and other living things as well), some foresters are planting their forests in ways that encourage short-eared owls, like this one, as a natural vole control.

Fit for the future

In short, we must learn to live sustainably, which means in ways that can continue forever without causing damage. At the Earth Summit in Rio de Janeiro, Brazil, in 1992, world governments signed a declaration promising to "achieve sustainable development and a higher quality of life for all people." Sustainable development means that we must obtain everything that we need today without spoiling the prospects for people in the future.

The previous chapters have shown that many things that governments and people do at present are completely unsustainable. We are smashing and grabbing from the planet, and that is robbing the future.

But did the governments of the world mean what they promised? Can they really change their ways? And can we?

◁ Flourishing living systems like this coral reef are a sign that the planet is healthy. When they start to fail, we know that the planet is ailing and that treatment is necessary.

GLOSSARY

adaptation Modification of the characteristics of an organism, through natural selection, in a way that increases its chances of survival in a particular environment.

aestivation The slowing down of the bodily activity of an animal during the summer period. Compare HIBERNATION.

alga A member of a group of plantlike photosynthetic organisms, including single-celled organisms and seaweeds.

animal An organism that can usually move freely, that relies on food from its surroundings, and whose cells are not surrounded by a cell wall.

bacteria A group of single-celled microscopic organisms, often shaped like spheres, rods, or spirals, that multiply by splitting in two. Most live by breaking down dead plant or animal material, but a few are parasites, causing infections in plants or animals.

biodiversity The total variety of living things, including the variety between individuals, between species, and within ecosystems.

biomass The weight of all living organisms in a given area.

bivalve An animal such as a mussel or a clam that has a two-piece, hinged shell.

broadleaves Trees such as oak and beech that have leaves in the form of flattened blades. Compare CONIFERS.

camouflage Coloration, shape, or behavior that allows a plant or animal to be hidden against its background.

carnivore An animal or plant that feeds on animal tissue (flesh).

cell One of the individual compartments of which all plant and animal tissue is composed, consisting of cytoplasm surrounded by a membrane and, in plants, also by a nonliving wall.

chlorophyll The chemical in green plants that drives photosynthesis.

cold-blooded A term used loosely for animals (fish, amphibians, and reptiles) that rely on outside environmental factors, such as the sun's heat, to keep their body temperature in the range essential for their survival. The strictly correct term is "poikilothermic." Compare WARM-BLOODED.

community A group of organisms that live together in a particular area or habitat and whose lives are to some extent interdependent.

conifers Trees that have needle-shaped leaves and whose seeds are usually produced in a cone. Compare BROADLEAVES.

conservation Work to protect and preserve plants, animals, their habitats, and the natural processes by which they live and die. The term is also used of measures that aim to reduce waste (e.g., of energy).

cytoplasm The living material making up the contents of cells around the nucleus.

deciduous Describing plants that regularly shed all their leaves together in one season of the year, usually winter. Compare EVERGREEN.

diapause A state in which the living processes of cold-blooded animals are temporarily slowed down in response to unfavorable environmental conditions. Compare HIBERNATION and AESTIVATION.

ecology The study of the interrelationship between plants, animals, and the environment in which they live.

ecosystem A natural, balanced living system produced by the interaction between organisms and their environment.

environment The surroundings of an organism, including physical factors and the effects of other organisms. The term is also used to refer to the combined physical processes and characteristics of the earth.

evergreen Describing plants that retain green leaves throughout the year. Compare DECIDUOUS.

evolution The natural process by which the characteristics of a group of organisms progressively change over many generations, in response to natural selection, resulting eventually in the appearance of new species.

extinction The disappearance, brought about by natural or unnatural means, of an entire species.

fertilizer A chemical or naturally occurring substance used to assist the growth of farmed plants.

fungus An organism belonging to a group that feeds by decomposing dead plant or animal matter. They vary from single cells to more complex organisms with fruiting bodies in the shape of mushrooms or toadstools.

genes Chemical messengers, passed from parent to offspring, that control the characteristics which the offspring inherit from their parents.

genetic Relating to genes and the process of inheritance.

germination The sprouting of a seed or spore as a new plant begins to grow.

global warming The perceived increase in average world temperatures as a result of pollution in the atmosphere.

greenhouse effect The natural mechanism by which the atmosphere traps heat close to the surface of the earth. The effect has been increased by excess levels of some gases as a result of pollution, leading to global warming.

habitat The type of locality, including all the environmental factors affecting it, in which a plant or animal is adapted to live.

herbivore An animal that eats leaves or other plant material.

hibernation The slowing down of the bodily processes of a warm-blooded animal to allow it to survive through a prolonged winter period. Compare AESTIVATION and DIAPAUSE.

host The animal or plant from which a parasite takes its food.

inheritance The process by which the characteristics of an individual are passed on to the next generation, as a result of the transmission of genes from parent to offspring.

invertebrates Animals without backbones.

larva The young stage in an animal's life cycle, which looks different to and lives differently from the adult animal.

lichen An organism resembling a stony crust, a leafy flap, or a beardlike tassel, formed by a close partnership between an alga and a fungus.

microbe A microscopic organism.

migration Regular two-way, seasonal movement of animals between separate living areas.

mineral A naturally occurring chemical substance derived from the rocks.

natural selection The process by which organisms best suited to living in an environment have the greatest chance of survival to breeding, thereby leading to progressive change in their characteristics.

nutrients Chemicals, obtained from various sources, that an organism needs for healthy living and growth.

nutrition The process of feeding and the subsequent absorption of the food material into the cells of an organism.

organism An individual living being.

ozone A naturally occurring poisonous gas, consisting of three oxygen atoms. Its presence in the upper atmosphere reduces levels of ultraviolet radiation reaching the surface of the planet.

parasite A plant or animal that takes its nutrition partly or wholly from another organism, called its host.

pesticide A poisonous chemical used to kill animals or plants regarded as pests by humans.

photosynthesis The process by which a plant makes food, in the form of sugars, from carbon dioxide and water, using the energy of sunlight, in its green leaves.

phytoplankton Tiny free-floating plant-like photosynthetic organisms in the plankton.

plankton Minute animals, plants, and other organisms floating freely in the waters of seas, rivers, ponds, and lakes.

plant An organism that makes its food by photosynthesis or is related to other organisms that do so; generally it will be unable to move between points under its own control and will have cells surrounded by a membrane and cell wall.

pollination The transfer of pollen between one flower and another by an animal, wind, or water, thus allowing the second flower to set seed.

pollution The damaging presence of abnormally high quantities of harmful substances in the environment, generally as a result of human activities.

predator An animal that lives by hunting other animals.

prey An animal killed and eaten by another animal.

primary producer An organism that manufactures food by photosynthesis.

Protista A group of organisms comprising single-celled animals, algae, bacteria, and related forms.

respiration The breaking down of food molecules in a cell to generate energy, using up oxygen and producing carbon dioxide as a waste gas.

ruminant A grazing animal whose stomach contains a special compartment called a rumen and that coughs up its partially digested food to "chew the cud."

saprophyte A plant or other organism that relies on dead plant material for its nutrition.

species A group of individual organisms that share common characteristics and that are capable of breeding with one another and passing these characteristics on to their offspring.

spore The equivalent of a seed in a lower plant, such as a fern or moss, consisting of a microscopic packet of a few cells, which can develop into a new plant in the right conditions.

stomata (singular **stoma**) Pores on the surface of a leaf allowing it to exchange gases and water vapor with the atmosphere around it.

subspecies An isolated group of individual organisms that has slightly different characteristics from similar organisms elsewhere.

succession A changing sequence of plant and animal communities occurring at one place over a period of time, resulting from changes in the environment brought about by these communities.

sustainability The process by which a resource can be carefully exploited today without damaging future supplies of the resource or the environment.

symbiosis An intimate partnership between two organisms in which the advantages to both partners outweigh the disadvantages to either.

tissue The massed cells forming the structure of an organism or of a particular part of the organism (e.g., nerve tissue, muscle tissue, etc.).

trace element A chemical substance required in minute quantities by an organism for healthy growth.

vertebrate An animal with a backbone.

warm-blooded A term used loosely for animals (birds and mammals) that are able to keep their body temperature constant by breaking down food or stored fat by respiration to produce heat. The strictly correct term is "homoiothermic." Compare COLD-BLOODED.

zonation The arrangement of zones of different plant and animal communities over a stretch of ground, in response to a gradation of environmental conditions.

zooplankton Tiny free-floating animals in the plankton, including the larvae of animals that live attached to rocks (etc.) as adults.

INDEX

ACKNOWLEDGMENTS

The author would like to thank his wife, Sue, for constant support and advice, especially in the *Kingdom of the Deep* chapter.

Design and typesetting: Getset Ltd

Picture research: Alex and Dora Goldberg, Image Select (London)

Abbreviations: t = top; b = bottom; l = left; r = right; c = center; back = background

Photographs
The publishers would like to thank the following for permission to reproduce the following photographs:

Acacia Productions: 129b both (Edward Milner)
Brian & Cherry Alexander: 46b; 69
Andromeda Oxford Limited: 131tr
Ardea: 24b (D&E Parer-Cook); 46r (Ian Beames); 108b (Liz & Tony Bomford); 123tl; 147br (John E. Swedberg)
Biofotos/Heather Angel: 4l (Jeremy Thomas); 13; 33 all (Jeremy Thomas); 39c; 43b; 52b (Andrew Henley); 138t
Bridgeman Art Library: 139t (Victoria & Albert Museum)
British Antarctic Survey: 60 (R. M. Laws); 62b (W. Block)
Britstock/IFA (London): 6–7 (Birgit Koch)
Michael Chinery: 80c
Bruce Coleman (UK Ltd): 1 (Michael Fogden); 30b (Quentin Bennett); 44r (John Shaw); 46tl (John Shaw); 102–103b (Michael Fogden); 141b (Norman Myers); 143br (Luiz Claudio Marigo)
Bruce Coleman (Inc) USA: 114t
Earth Scenes: 95bl (Patti Murray)
Christer Fredriksson: 82
Gamma Presse: 9t main (Patrick Aventurier); 147b inset; 149tl (Bartholomew); 149b (NASA/LIAISON); 150b (Eric Sander); 151 (A Ribeiro)
Greenpeace: 147t (Merjenburgh); 150–151 (Culley)
Holt Studios: 9t inset (Nigel Catlin)
Images Colour Library: 146b main
Image Select/NASA: 152 main
Images of Africa Photobank: 8c (David Keith Jones)
Jacana Scientific Control (Paris): 2 (François Gohier); 3 (Jean-Paul Ferrero); 4r (Yves Lanceau); 5t (G. Loreryl); 5bl (François Gohier); 10b (Sophie de Wilde); 12r (K. Ross); 18 (Anup Shah); 19t; 24t (Claude Pissavini); 31t; 35c (Pierre Pilloud); 36b (Bill & Peter Boyle); 37t (Jean-Philippe Varin); 37b (François Gohier); 38b (Serge Pecolatto); 38r (Walter Geiersperger); 40t (G. C. Kelley); 40b (Jean-Paul Ferrero); 41r (C. & M. Moiton); 43t (D. Guravich); 45b (Stephen Krasemann); 48 (François Gohier); 49t (John Cancalosi); 49bl (J. F. Hellio & N. Van Ingen); 51l (François Gohier); 51r (G. Lorenz); 52t (Alain Devez); 54–55 (Yann Arthus-Bertrand); 56 (Jean-Paul Ferrero) 57t; 63b; 63cr inset (Suinot); 64t (D. Guravich); 64 inset (W. Bacon); 64–65 (Sylvain Cordier); 65b (Brian Hawkes); 67t (J. Lepore); 68; 701

(J. P. Saussez); 75 (Pierre Pilloud); 79b (Sylvain Cordier); 80t (Rudolf Konig); 81t inset; 83t (A. Carrara); 84t (François Gohier); 84bl (Sylvain Cordier); 84br; 86b (François Gohier); 87t (Gunter Ziesler); 87b (Clem Haagner); 88cl; 88–89 (Jean Dragesco); 89tl (Jean-Paul Hervy); 89tr (Werner Layer); 89b (Brian Hawkes); 91b (A. Dubourg); 92t (Yves Lanceau); 92b (A. R. Devez); 93t (John Cancalosi); 94t (T McHugh); 95br (R. Konig); 97 (François Gohier); 100bl (Ferrero/Labat); 103r (M. McCoy); 105b (John Cancalosi); 108–109b (M. Loup); 109b (J. F. Hellio & N. Van Ingen); 110c (François Gohier); 117t (Jean-Marie Bassot); 117b (Herve Chaumenton); 121b main (Armelle Kerneis-Dragesco); 124t (Elizabeth Lemoine); 125b (Jean-Louis Dubois); 126t (Ferrero-Labat); 126bl; 126bc (Herve Chaumeton); 127tr (Jacques Brun); 128 (François Gohier); 129tr (Sophie de Wilde); 130tl (D. Cande); 130tr (G. Moon); 134t (Jean Claude Pissavini); 135 inset (G. Sommer); 139b (Paul A. Zahl); 141t (François Gohier); 142b; 144b (Alain Le Toquin); 146 inset (Elizabeth Lemoine)
Andrew Kitchener, Royal Museum of Scotland, Edinburgh: 9br
Frank Lane Picture Library: 91t (R. S. Chundawat); 125t (Tony Wharton)
Natural History Photographic Agency: 10r (John Shaw); 26 (M. I. Walker); 32 (Peter Parks); 46b (Brian & Cherry Alexander); 57c (Michael Leach); 62tl (Peter Johnson); 104t (Ivan Polunin); 106 (Stephen Dalton); 112 (Peter Parks); 114bl (Bill Wood); 131bl; 131br (Laurie Campbell); 135t (Michael Leach); 144t (Alan Williams); 152l inset (G. I. Bernard)
Nature Photographers Limited: 101 (Paul Sterry)
Jennifer Owen: 134c main
Oxford Scientific Films: 5br (Anthony Bannister); 11t (Dr. C. E. Jeffree); 12r (K. Ross); 14 (Tim Shepherd); 17r; 17br (Kathie Atkinson); 17bl (Michael Leach); 19b (Mark Hamblin); 23b (Kathie Atkinson); 28bl; 28br (G. I. Bernard); 29t (Muzz Murray); 29b both (G. I. Bernard); 34–35 (Sinclair Stammers); 36t (Judd Cooney); 38t (Mike Birkhead); 41l (Guari Thurston); 45t (Konrad Wothe); 49r (Michael Fogden); 50bl (Fran Allan); 50br (Sinclair Stammers); 61t (Ben Osborne); 62tr (Doug Allan); 65tr (Tom McHugh); 66bl (Martyn Chillmaid); 66br (Owen Newman); 71b (Tim Davis/Photo Research Inc); 79t (G. A. Maclean); 80t (R. Konig); 81c main (Stuart Bebb); 83b (Gordon Maclean); 85tc (Joe MacDonald); 88bc (Anthony Bannister); 90 (Anthony Bannister); 93b (Michael Fogden); 94b (Raymond A. Mendez); 96b (Roger Jackman); 98 (Harold Taylor); 101b (Peter Gould); 102tl (Michael Fogden); 102tr (Wendy Shattil & Bob Rozinski); 108–109b (John Paling); 114br (Harry Taylor); 119b (Howard Hall); 121b inset (Barrie Watts); 132t (John Paling); 133b (Kathie Atkinson); 134bl (Alan and Sandy Carey); 136–137 (Richard Packwood); 142t (Tony Martin); 145b (Richard Packwood); 142tl (Tony Martin); 145b (Richard Packwood); 153r (Mike Birkhead)
Klaus Paysan (Stuttgart): 72, 105t

Planet Earth Pictures: 8br (Anup & Mahos Shah); 9c (Richard Matthews); 12 (Mark Conlin); 20 (Geoff du Feu); 22 (Richard Chesher); 23t (Robert Hessler); 27tr (John Downer); 34bl (Peter Gasson); 35bl inset (W. B. Irwin); 35br inset (W. B. Irwin); 39t (Geoff du Feu); 42 (Franz J Camenzind); 44l (Cameron Read); 47 (Norbert Wu); 50tl (Pete Atkinson); 61b (Norman Cobley); 63t (Peter Scoones); 66tl (James King); 70r (Keith Scholey); 71t (John Downer); 73 (Robert A. Jureit); 76t (J. M. Heap); 76b (Richard Matthews); 80b (Ernest Neal); 81b inset (John Lythgoe); 83cl; 83cr (Hans C. Heap); 85cr (Franz J. Camenzind); 86t (John Waters); 88t (Ken Lucas); 88bl inset (John Downer); 95t (Georgette Douwma); 96t (Roger de la Harpe); 100t (André Bartschi); 100br (Brian Kenney); 101tc (John Bracegirdle); 104b (André Bartschi); 110b inset (Peter Scoones); 113t (Robert Arnold); 113cl (Christian Petron); 113br (Doug Perrine); 115t (Peter Scoones); 116bl (Colin Doeg); 118 (Gary Bell); 123tr (Jim Greenfield); 126br (Keith Scholey); 132b (Peter Chippendale); 133t (Gordon James); 134c inset (Wayne Harris); 140t (Jeannie MacKinnon); 140–141 (Jonathan Scott); 142–143t (Marty Snyderman); 146t (Warren Williams); 153l (Peter Scoones)
Rex Features: 149r
Ann Ronan/Image Select (London) 66cr; 130b; 131tl; 138b
Science Photo Library: 11r inset
Michael Scott: 10t; 16; 17tl; 27b; 34tr; 57b; 67b; 74–75 main; 74 inset; 109t; 109c; 122; 123c; 124b
Sue Scott: 30t; 39r inset; 115b; 116t; 116br; 119c; 120c; 120br; 120bl; 121t
Survival Anglia: 102bl (Dieter & Mary Plage)
Windrush Photos: 54t (David M. Cottridge)
WWF (Gland Switzerland): 11b main; 15; 28t (M. Harvey); 31b; 101t inset (R. Seitre); 108t inset (John Newby); 152r inset (Mark Edwards)
Zefa: 74; 140b

Illustrations and diagrams
Graham Allen: 25cr
John Barber: 76–77; 80–81; 107
Brian Beckett: 120t
Stefan Chabluk: 47; 55br; 69; 122; 145; 148
Nick Hawken: 8t; 18; 26tc; 42; 48; 58–59; 60; 82; 144; 150
Oxford Illustrators: 20; 22; 25t; 25b; 53c; 54tr; 54bl; 56; 78t; 78c; 85; 90–91; 99b; 112; 115; 119; 128–129; 147
David More: 14c; 14br; 15; 16; 29 inset; 34 inset; 72; 120; 123; 125
Michael Strand: 21
Peter Visscher: 8b; 13; 23; 26–27c; 53t; 53b; 56tl; 56bl; 78b; 99t; 111; 127
Michael Woods: 12t; 34; 52; 68; 90tc; 98; 106; 110; 124; 128tc; 132; 138

Cover
Front main picture: Science Photo Library (Claude Nuridsany and Marie Perennou)
Front main picture background: Letraset Phototone
Back main picture: Oxford Scientific Films (Mike Birkhead)
Background for front and back: Zefa (Frans Lanting)